Prioritizing Water Main Replacement and Rehabilitation

The mission of the Awwa Research Foundation is to advance the science of water to improve the quality of life. Funded primarily through annual subscription payments from over 1,000 utilities, consulting firms, and manufacturers in North America and abroad, AwwaRF sponsors research on all aspects of drinking water, including supply and resources, treatment, monitoring and analysis, distribution, management, and health effects.

From its headquarters in Denver, Colorado, the AwwaRF staff directs and supports the efforts of over 500 volunteers, who are the heart of the research program. These volunteers, serving on various boards and committees, use their expertise to select and monitor research studies to benefit the entire drinking water community.

Research findings are disseminated through a number of technology transfer activities, including research reports, conferences, videotape summaries, and periodicals.

Prioritizing Water Main Replacement and Rehabilitation

Prepared by:
Arun K. Deb, Frank M. Grablutz, Yakir J. Hasit,
and **Jerry K. Snyder**
Roy F. Weston, Inc.
West Chester, PA 19380

and

G.V. Loganathan and **Newland Agbenowsi**
Virginia Polytechnic Institute and State University
Blacksburg, VA 24061

Sponsored by:
Awwa Research Foundation
6666 West Quincy Avenue
Denver, CO 80235-3098

Published by the
Awwa Research Foundation and
American Water Works Association

Disclaimer

This study was funded by the Awwa Research Foundation (AwwaRF). AwwaRF assumes no responsibility for the content of the research study reported in this publication or the opinions or statements of fact expressed in the report. The mention of trade names for commercial products does not represent or imply the approval or endorsement of AwwaRF. This report is presented solely for informational purposes.

Library of Congress Cataloging-in-Publication Data has been applied for.

Printed in the U.S.A.

ISBN 1-58321-216-7

Printed on recycled paper

CONTENTS

TABLES

FIGURES

FOREWORD

The Awwa Research Foundation is a nonprofit corporation that is dedicated to the implementation of a research effort to help utilities respond to regulatory requirements and traditional high-priority concerns of the industry. The research agenda is developed through a process of consultation with subscribers and drinking water professionals. Under the umbrella of a Strategic Research Plan, the Research Advisory Council prioritizes the suggested projects based upon current and future needs, applicability, and past work; the recommendations are forwarded to the Board of Trustees for final selection. The foundation also sponsors research projects through the unsolicited proposal process; the Collaborative Research, Research Applications, and Tailored Collaboration programs; and various joint research efforts with organizations such as the U.S. Environmental Protection Agency, the U.S. Bureau of Reclamation, and the Association of California Water Agencies.

This publication is a result of one of those sponsored studies, and it is hoped that its findings will be applied in communities throughout the world. The following report serves not only as a means of communicating the results of the water industry's centralized research program but also as a tool to enlist the further support of the nonmember utilities and individuals.

Projects are managed closely from their inception to the final report by the foundation's staff and large cadre of volunteers who willingly contribute their time and expertise. The foundation serves a planning and management function and awards contracts to other institutions such as water utilities, universities, and engineering firms. The funding for this research effort comes primarily from the Subscription Program, through which water utilities subscribe to the research program and make an annual payment proportionate to the volume of water they deliver and consultants and manufacturers subscribe based on their annual billings. The program offers a cost-effective and fair method for funding research in the public interest.

A broad spectrum of water supply issues is addressed by the foundation's research agenda: resources, treatment and operations, distribution and storage, water quality and analysis, toxicology, economics, and management. The ultimate purpose of the coordinated effort is to assist water suppliers to provide the highest possible quality of water economically and reliably. The true benefits are realized when the results are implemented at the utility level. The foundation's trustees are pleased to offer this publication as a contribution toward that end.

A large portion of a water utility's investment is dedicated to their distribution system. Yet, the condition of the distribution system, once put in place, is generally unknown. Often times the only opportunity available to assess a water main's condition is during a main break or scheduled repair. However to date no standardized methodologies exist to establish the proper data and sample collection to be taken. This project aims at providing guidance and methodologies for the collection of relevant post-mortem field data and field samples after water main leaks and failures. The data collected is used in a predictive distribution system, condition assessment model to target, evaluate, and prioritize infrastructure rehabilitation investments.

Edmund G. Archuleta, P.E.
Chair, Board of Trustees
Awwa Research Foundation

James F. Manwaring, P.E.
Executive Director
Awwa Research Foundation

ACKNOWLEDGMENTS

The authors of this report thank the Awwa Research Foundation (AwwaRF) for its support of this study, and the AwwaRF project manager Stephanie Morales for her help and guidance in the execution of the project. The authors acknowledge the input provided by the members of the project advisory committee (PAC), namely Tony Woodward, Thames Water Utilities, UK, Ken Mueller, St. Louis County Water Company (SLCWC), Mo., and Richard Nelson, Black and Veatch, Corporation, Kansas City, Mo.

The key aspect of this study was the collection of pipe break and related data from participating water utilities. The utilities spent considerable amount of their resources in the collection and testing of pipe and soil samples. The key individuals at these utilities were Fattah Hashem-Zadeh, Regional Municipality of Ottawa-Carleton (RMOC), Ont., John Dyksen, United Water New Jersey (UWNJ), N.J., Joe Thurwanger, Philadelphia Suburban Water Company (PSWC), Pa., Yone Akagi, City of Portland Bureau of Water (PBW), Ore., and Tim Dennis, City of Toronto Public Works, Ont. Considerable information on material testing was provided by Edgar Navera, Philadelphia Water Department (PWD), Pa. Similarly, main break databases were provided by PWD, SLCWC, PSWC, RMOC, UWNJ, and Ft. Worth Water Department.

In addition to the authors, other project team members included Helena Alegre, Laboratorio Nacional de Engenharia Civil (LNEC), (i.e. Civil Engineering National Laboratory), Portugal, Ronald von Autenried, Buck, Seifert & Jost, Inc. (BS&J), N.J., and Donald Ballantyne, EQE International, Wash. Acknowledgement is also due to Balvant Rajani from National Research Council Canada, Ont., who, as the principal investigator of a separate but related AwwaRF project, provided information on water main breaks.

Input was also provided by the participants of two project workshops held in January and October 1998. Excluding the individuals already acknowledged above, workshop participants included Lloyd Miller, PBW, Daniel Moe, Denver Water, Colo., Charles Zitomer, Dennis Blair, Stephan Furtek, Tom Kulesza, Philip Bobb-Semple and Ahmad Samadi, PWD, George Rizzo, Frederick MacMillan and Leonard Hotham, U.S. EPA Region III, Pa., David Hughes, PSWC, Pen Tao, UWNJ, Van Speight, Showri Nandagiri and Harish Jajoo, City of Houston Department of Public Works and Engineering, Texas, Laurence Murphy, United Water New Rochelle, N.Y., Ivan Tasky, United Water Jersey City, N.J., Curtis Cochrane and Bernard Gascon, District of

Columbia Water and Sewer Authority, Guido von Autonreid, BS&J, and Gregory Welter, O'Brien and Gere Engineers, Md.

EXECUTIVE SUMMARY

INTRODUCTION

A key factor in developing water main rehabilitation and replacement programs (collectively referred to as "renewal" programs) is to know their physical condition. Main breaks are therefore a primary source of data for the condition of pipes, because when pipes are excavated for repair they provide the opportunity for data collection. Furthermore, analyzing trends in main breaks, associating physical characteristics shared by similar groups of pipes in the system, and considering spatial trends in main breaks also supports decision making in main rehabilitation and replacement.

The focus of this project was the collection and application of data associated with water main breaks. Specifically, the objectives of the project were to:

- Provide guidance and develop methodologies for collecting main break field data and pertinent field samples for laboratory analysis.
- Develop procedures for data analyses to assess type of failure, reason for failure, and the condition of water mains.
- Describe procedures to use these data to target, evaluate, and prioritize infrastructure rehabilitation and replacement investments.
- Develop a modeling system incorporating these data to predict and prioritize infrastructure rehabilitation and replacement programs.

That is, the ultimate goal of the project was to describe procedures and tools that a utility could use in developing practical and cost-effective distribution system renewal programs by taking advantage of the data collected during water main breaks. A primary focus of the project was to make sure that the data requirements were practical and within reach of most utilities.

The project team was led by Roy F. Weston, Inc. (WESTON), West Chester, Pa., and included Virginia Polytechnic Institute (VA Tech), Blacksburg, Va., Laboratorio Nacional de Engenharia Civil (LNEC), Lisbon, Portugal, i.e. the Civil Engineering National Laboratory of Portugal, Buck, Seifert & Jost, Inc. (BS&J), Paramus, N.J., and EQE International, Seattle, Wash.

REVIEW OF CURRENT INFORMATION

Numerous sources were utilized to compile background information on main breaks and decision criteria for water main renewal. These sources included published literature by water utilities, academic sources and gas industry, AWWA's WATER:\STATS database and a questionnaire survey

The questionnaire survey found that between 1993 and 1997 the average break rate in North America was between 21 and 25/breaks/100 mi/yr, while the average break rate in Europe was higher with about 80 breaks/100 mi/yr. Less than half of the North American utilities were implementing formal programs for controlling main breaks, while 80% of the European utilities had such programs. Furthermore, 85% of European utilities had their main break data computerized versus 70% of North American utilities. The type of main break information collected, such as date, address, time, temperature, etc., was similar in North American and European utilities.

To aid the investigation, water main break databases were obtained from six utilities. The information was representative of the wide range of detail typical of such databases. The simplest database recorded only the date, address, pipe material, pipe diameter, break type, and repair type. Conversely, the most detailed database included the same data as well as soil conditions, labor times, equipment used, and other useful fields. Even the simplest databases contained information found to be useful for making decisions about water main renewal.

A literature search identified various methods used to prioritize the renewal of water mains. These can be grouped into the following four categories:

- Deterioration point assignment methods: use a scoring system to assign points to pipes based on various characteristics of the pipe and its environment
- Break-even analyses: use economics to compare the costs of repair versus replacement
- Failure probability and regression methods: predict the probability of future failures
- Mechanistic models which attempt to simulate the deterioration of a pipe over time and the loads to which a pipe is subjected. These mechanistic models require very

detailed pipe and environmental data, or must rely on general assumptions where the data are not available.

A review of gas utility practices showed that they often apply the concepts of risk management in developing infrastructure renewal programs. For the gas industry, risk management involves balancing the consequences of a gas main or pipeline failure against the costs of inspecting, maintaining, rehabilitating, or replacing the pipes in question. Factors considered in estimating the probability of failure for gas mains include external and internal corrosion, third party damage, stress-corrosion cracking, material factors, construction factors, outside sources, and electrical disturbances. These factors are also applicable to pipes in a water distribution system.

WATER MAIN BREAK DATA COLLECTION

Various types of useful data can be collected every time a water main fails. These can be categorized as field data, office data, laboratory and field test data, and cost data. The types of data in each category include the following:

- Field data - the primary information about the failed pipe and its surroundings. These include:
  - Date of excavation
  - Name of employee filling out form
  - Address, nearest cross street, distance from cross street
  - Pipe material, diameter, cathodic protection, joint type
  - Condition of pipe interior and exterior
  - Bedding
  - Surface and traffic
  - Pipe and frost depth
  - Type, location and probable cause of failure
  - If pipe failures are caused by bedding erosion, the unsupported span length
  - A sketch of the pipe and the location of the failure.

- Office data - background information about the pipe. These include:
  - Name of employee filling out form
  - Water main ID, installation year, length
  - Water main break ID, and dates of previous breaks on the same water main ID
  - Typical pressure and flow in area of break
  - Water and air temperature, and number of consecutive days below 32°F
- Laboratory and field test data of pipes and soils which include:
  - Soil category, temperature, pH, moisture content, resistivity
  - Pipe wall thickness, modulus of rupture, fracture toughness
- Cost data which include:
  - Crew size and time on site
  - Interruption of water service to customers, estimated number of customers whose service was interrupted, and estimated time without service
  - Property damage

A guidance manual and forms incorporating the data listed above were used by participating utilities to record water main break events over an approximately 6-month long period. Based on their experiences, the guidance manual and the form were revised, and are provided in this report.

APPLICATIONS OF WATER MAIN BREAK DATA

Background

The applications of main break data vary from the simple data analysis to use of sophisticated mathematical models. The application depends on the types of data collected and the availability of resources at the utility. Regardless, even simple databases can be useful in the development of water main rehabilitation and replacement programs, such as keeping track of the maintenance history of the pipe.

Similarly, availability of historic water main break data allows a utility to analyze trends in main break occurrences. One type of trend analysis is the spatial distribution of water main breaks. Some utilities simply manually plot the locations of main breaks on a map. A similar but more sophisticated approach utilizes a GIS to track main break occurrences. For this type of analysis, the only type of water main break data that is required is the location of the break.

A review of water main breaks over time (i.e. temporal analysis) is another simple analysis that can be performed if data are collected and maintained over an extended period.

Mathematical models can be useful when numerous factors need to be incorporated into the development of progressive programs for the rehabilitation and replacement of pipes. Factors that can be incorporated in mathematical models can include earth load, truck/live load, working pressure, water hammer given certain pipe materials and thickness. Other factors include frost load, temperature variations in the soil-pipe-water environment, pipe bed condition, and corrosion.

Modeling

In this project a mechanistic model of the corrosion process and loss of strength of the pipe was developed to prioritize the replacement and rehabilitation of pipes. This mechanistic model consists of several modules. The loads that the pipe is subject to and the resulting material stresses are calculated by the Pipe Load Module (PLM). The corrosion process and the resulting loss of residual strength of the pipe are calculated by the Pipe Deterioration Module (PDM). The Statistical Correlation Module (SCM) calculates the residual strength of the pipe based upon the reduced wall thickness calculated by the PDM. The Pipe Break Module (PBM) compares the stresses on the pipe to the residual strength of the pipe. The ratio between these values represents a Safety Factor (SF) for the vulnerability of the pipe to failure. The model could be calibrated by calculating SF using the PBM for pipes listed in the utility's water main break database and confirming failure modes. A prioritized replacement plan can be developed by ranking pipes according to SF. Pipes with the lowest SF are most vulnerable to failure and should be addressed first.

MODELING CASE STUDY

Results

The mechanistic model developed in this project was tested with Regional Municipality of Ottawa-Carleton's (RMOC's) data to determine the applicability of the models in prioritizing main replacement programs. The data were gathered both from the RMOC water main inventory and the RMOC main break database. Assumptions were made when the required data were not available.

In evaluating the model, the safety factors (SF) for the combination of hoop and ring stresses (SF_{h-r}) and flexural and longitudinal stresses (SF_{f-l}) were considered. The average SFs were found to be 1.64 for both SF_{f-l} and SF_{h-r}. As expected, the less corrosive soils were found to have generally higher SFs than corrosive soils. The effect of pipe age on the SF was also examined. As expected, the SF was found to decrease over time.

Conclusions

The mechanistic model was able to predict changes in the SFs of cast iron pipe based upon age, soil type, and year of manufacture. Soil type was found to have the most pronounced effect on the predicted SF for the pipes. Also a greater variability was found in the predicted SFs for combined ring and hoop stress than for combined flexural and longitudinal stress.

RECOMMENDATIONS

Water main breaks provide an excellent opportunity for water utilities to gather information on the condition of pipes in their system. Data collection programs, however, must be coordinated with field crews in order for them to be effective. Furthermore, it is critical that collected data should be maintained in databases. Even the simplest databases might provide valuable insight in developing main rehabilitation and replacement programs. Essentially, the break data collected and stored should consist of field, office, laboratory testing, and cost. These include date, address, pipe material and age, pipe diameter, break type, repair type, soil conditions, crew time, etc. Depending on the extent of data collected, these data can then be used to conduct simple trend analyses or sophisticated mechanistic model analysis to plan for water main renewal. The mechanistic model can be used to prioritize water main renewal programs.

CHAPTER 1
INTRODUCTION

BACKGROUND AND OBJECTIVES

The principal link between a water utility and its customers is the distribution system. The customers' perceptions about the utility are often based upon service issues that arise in the distribution system. For example, interruptions in service due to main breaks may impact reliability of the system and can lead to customer dissatisfaction. Water quality problems created by the internal condition of pipes are generally aesthetic in nature, but similarly can lead to customer dissatisfaction and complaints. These types of problems are commonly a function of an aging pipe network. Because few utilities have actively pursued aggressive water main renewal (rehabilitation or replacement) programs, water utilities are now faced with the problem of renewing distribution systems that may have received little attention since the pipes were first installed. It has been estimated that between $77 and $325 billion (USEPA 1997, AWWA 1999) are needed to renew water distribution systems in the US over the next 25 years. Given the magnitude of this investment, water utilities need to develop practical and cost-effective programs for conducting this renewal.

A key factor in designing a renewal program is to understand the condition of pipes in the system. After all, it is the condition of these pipes that create the reliability and water quality problems that must be addressed. Although on-site, real-time inspection of a pipe is the ideal method to analyze the condition of pipes, this approach is expensive and cannot be practically applied to the entire distribution system. Non-destructive testing procedures are being developed that could allow for a more thorough evaluation of the distribution system, but these technologies are still evolving. Therefore, water utilities must take advantage of all opportunities for collecting data on the condition of pipes, develop standardized procedures for collecting the data, and identify and implement opportunities to put the data to use.

Although they are a source of frustration to both the utility and its customers, water main breaks provide an opportunity to evaluate the condition of pipes in the distribution system. Pipes are rarely exposed except when they are excavated to repair a main break. This provides an opportunity to record the type and possible cause of the break, observe the condition of the pipe exterior, collect information on the bedding and soil conditions, and to assess the condition of corrosion protection measures, if any. If a section of the pipe is removed in order to complete the repair, additional information related to the pipe interior can be gathered.

Water main breaks provide other opportunities for assessing the condition of the distribution system beyond the immediate physical data that is collected. For example, analyzing trends in main breaks, associating physical characteristics shared by similar groups of pipes in the system, and considering spatial trends in main breaks are only possible if sufficient data are collected over time.

The American Water Works Association Research Foundation (AWWARF) funded this project (Project 459) in order to assist utilities in the development of proactive water main rehabilitation and replacement programs. The focus of the project was on the collection and application of data associated with water main breaks. Specifically, the objectives of the project were to:

- Provide guidance and develop methodologies for collecting main break field data and pertinent field samples for laboratory analysis
- Develop procedures for data analyses to assess type of failure, reason for failure, and the condition of water mains
- Describe procedures to use these data to target, evaluate, and prioritize infrastructure rehabilitation and replacement investments
- Develop a modeling system incorporating these data to predict and prioritize infrastructure rehabilitation and replacement programs

As the project progressed, it became apparent that the major concern for most utilities was related to cast iron pipes. This is primarily due to the fact that cast iron pipes represent about half of the distribution system piping in North America (AWWA 1998). Cast iron pipes are also typically the oldest pipes in the distribution system and have the longest history of water main breaks. Therefore, this study mainly addresses cast iron pipes based on the data availability and the fact that water main renewal programs for the foreseeable future are likely to focus on this category of pipes. Water main break data collection practices and applications of main break data are applicable to any type of pipe material. The mechanistic modeling intended to help utilities prioritize pipes for renewal is applicable only to cast iron pipes, however.

PROJECT APPROACH

The project included a review of how utilities presently collect data during water main breaks, and an investigation of how utilities apply or use the data. Recommendations were made regarding the types of data that should be collected, and these recommendations were tested in the field by several utilities. The ultimate goal of the project was to describe procedures and tools that a utility could use in developing practical and cost-effective distribution system renewal programs by taking advantage of the data collected during water main breaks. A primary focus of the project was to make sure that the data requirements were practical and within reach of most utilities.

The project included the following tasks:

- A review and evaluation of water main break data collection practices and pipe renewal methodologies was conducted. This included a review of gas utility practices that are applicable to the water industry.
- A questionnaire survey was mailed to North American and European water utilities to identify current main break data collection practices. Selected utilities were contacted to provide additional information, including actual databases of main breaks and associated data.
- Draft guidance that provided procedures for data collection related to water main breaks was developed.
- An expert workshop was held in which the project team, the project advisory committee (PAC), and utilities discussed practicality issues associated with main break data collection. The proposed mechanistic modeling approach was also discussed.
- The draft data collection guidance was revised on the basis of workshop inputs. The modeling approach was modified to better conform with the expected data availability.
- The data collection guidance was tested in a field data collection effort.
- The data collection guidance was modified and finalized based on the field test.
- The mechanistic model was completed and applied to test case utilities.

- Findings and recommendations were developed and presented in this report.

PROJECT TEAM

The project team included Roy F. Weston, Inc. (WESTON), West Chester, Pa., Virginia Polytechnic Institute and State University (VA Tech), Blacksburg, Va., the Civil Engineering National Laboratory of Portugal (LNEC), Lisbon, Portugal, Buck, Seifert & Jost, Inc. (BS&J), Paramus, N.J., and EQE International, Seattle, Wash. WESTON was the prime contractor and led the overall effort. VA Tech investigated existing main break predictive models and developed the mechanistic model used in this project. LNEC was responsible for data collection from European water utilities, and provided insight into European practices. BS&J provided the input on actual field application of data collection practices related to water main breaks, in particular to large main breaks. They also prepared Appendix A that focuses on large water main breaks. EQE International provided input and guidance on seismic issues related to water main breaks, and on data collection requirements specific to seismic issues.

REPORT ORGANIZATION

This chapter provides the background and the objectives of the project. Chapter 2 presents a review of available literature and actual utility practices related to the collection, management, and analysis of water main break information. The literature review includes an examination of gas utility practices, given the similarities between gas and water utility distribution systems. Actual utility practices were examined both from the literature review as well as from the questionnaire responses received from about seventy utilities throughout North America and Europe.

Data needs related to water main breaks are described in Chapter 3. This chapter also describes the data collection procedures required to satisfy the modeling needs, and describes the field case studies for utilities that participated in a field test application of the proposed data collection procedures and requirements.

Applications using the main break data are presented in Chapter 4. It is recognized that utilities may have different levels of resources available for the collection and use of water main break data. Therefore, several useful applications are presented, ranging from simple analyses of main break data to a mechanistic model that can prioritize cast iron water mains for replacement based upon their vulnerability to failure. Chapter 5 describes the application of this model to a test case utility. The process of data collection, data input, and running of the models is described. Problems encountered and limitations of the model are presented along with the results of the modeling. Finally, Chapter 6 summarizes the major findings and recommendations of the study. This includes recommendations on data collection procedures and findings on the practicality of data requirements for the modeling.

CHAPTER 2
REVIEW OF CURRENT INFORMATION

This chapter summarizes the background information collected as part of this study. A variety of sources were utilized to compile the information. A search was conducted to identify published literature addressing water main break data collection practices and water main renewal prioritization methodologies. The search also included gas utilities that face similar challenges in maintaining piping distribution systems. Insights obtained from AWWA's WATER:\STATS database and from a questionnaire survey conducted for this project are also described in this chapter.

UTILITY MAIN BREAK DATA COLLECTION STUDIES

One of the most detailed studies undertaken to assess the physical condition of water mains subject to failure was performed by the Philadelphia Water Department (PWD) Materials Testing Laboratory (MTL) under their broken water main sampling program (PWD 1985a). The specific objectives of this program were to:

1. Document the structural deterioration of water mains using standard mechanical strength tests
2. Do an overall inspection of the water mains that have failed
3. Conduct a pipe-wall analysis of the mains
4. Classify the soils surrounding the mains

Three types of test were conducted on cast iron pipes. The first one included a visual inspection and thickness analysis to determine the degree and extent of corrosion in pipes. The test was conducted on cross-sectional rings cut from damaged pipes. The second type of test included a series of mechanical tests to determine the structural integrity of pipes. The tests performed were Brinell hardness (using hardness squares), modulus of rupture (using talbot strips), tensile strength (using tensile bars) and percent carbon (using percent carbon chips) tests. The first three of the mechanical tests were for determining the main's ability to withstand a variety of stresses without fracture, while the last one was for determining the remaining chemical structure of the pipe. The third type of test involved soil analyses to determine the impact of the surrounding soils on the physical integrity of the pipes. The analyses included soil classification and soil resistivity tests.

The visual and thickness analyses found that the corrosion problems were primarily tuberculation and external graphitization. The tensile strength tests indicated that most of the samples had the same tensile strength as their design values. The modulus of rupture tests showed that the majority had one sixth of their original strength. In the Brinell hardness test, the older pipes were unexpectedly found to have higher hardness numbers. This was attributed to their higher carbon contents, which makes iron harder but more brittle. The percent carbon tests results were in the range of 3%-4%.

In an accompanying study, PWD developed a computer-based model to assess the structural condition of the water mains in the city (PWD 1985a). As most of the pipes in the system were cast iron, the model focused on this type of pipe. The data needed to develop the model was partially collected under the accompanying study (PWD 1985b). The factors considered in the development of the model included:

- Pipe characteristics such as diameter and effective wall thickness
- Structural properties such as bursting tensile strength, modulus of rupture and tensile strength
- Longitudinal and transverse forces and stresses such as bending stress, thermal contraction stress, longitudinal pressure stress, hoop stress, and ring stress
- External loads such as earth load, truck superloads, frost loads
- Age related factors such as cast iron manufacturing techniques, cast iron strength of main, design practices at time of construction, construction practices and materials at installation and deterioration due to corrosion
- Corrosion, including internal corrosion, external galvanic corrosion and stray direct current
- Soil characteristics such as soil classification, soil moisture, soil resistivity

Based on these factors, the model generated a "structural conditions rating", where values above 1.0 were classified as satisfactory, and values below 1.0 were classified as "questionable". A sensitivity analysis of the various factors showed that the most sensitive factor was beam span length, followed by internal corrosion rate, diameter, temperature range, external corrosion rate and working pressure. Thus, it was concluded that the success of the model depended on the accuracy of determining these difficult to quantify factors.

In another main break study, Severn Trent Water (U.K.) analyzed the main break information in its databases to have a better understanding of main breaks within the overall water system infrastructure (Kane 1996). It concluded that:

- Most of the main breaks occur in cast iron pipes.
- The break rate among cleaned and lined cast iron pipes is about a quarter of the break rate of unlined cast iron pipes. Thus, cleaning and lining is recommended for structurally sound unlined cast iron pipes.
- Break rates in corrosive soils are double the rates of pipes in non-corrosive soils.
- Break rate is 50% higher in soils that expand and contract due to soil moisture.
- Cold weather affects break rates both by the duration and the severity of cold weather.
- Most main breaks are circumferential breaks, especially for cast iron, asbestos cement, and medium density polyethylene (MDPE) pipes.
- The highest numbers of breaks occur in winter months.

In a separate study describing the water main renewal methodology of Severn Trent Water, Kane discusses some of the failure mechanisms of water mains, based on pipe material (Kane 1996). Tuberculation and the impact of corrosion on the residual strength of pipes are included in his discussion.

AWWA WATER:\STATS DATABASE

The WATER:\STATS database was compiled as a joint effort by AWWA and AWWARF. The latest survey of water utilities was conducted in 1996 and 898 utilities responded. Data are available in the areas of general information, utility revenue, treatment practices, finances, distribution systems, and water quality.

WATER:\STATS provides the total length of pipe in each system, the length of pipe replaced in the previous year, the length of pipe extensions for the previous year, a breakdown of pipe length by material, and the number of main breaks for the previous five years. The database shows that cast iron and ductile iron pipes are predominant in both the United States and Canada. In the U.S. cast iron pipe accounts for 48% of the total distribution system piping, compared to 44% in Canada. For ductile iron the percentages are 19% in the U.S. and 25% in Canada. Thus, cast iron and ductile iron together account for about two-thirds of the pipe in both countries.

The WATER:\STATS data also show that the 702 utilities that provided information on water main breaks experienced 90,952 breaks in 1995. In other terms, this is equivalent to approximately 23 breaks/100 mi/yr for the utilities that responded to the WATER:\STATS survey.

QUESTIONNAIRE SURVEY OF WATER UTILITIES

A questionnaire survey of North American and European water utilities was conducted. The primary objective was to identify utilities that have main break inventories and related information in computerized databases. A secondary objective was to analyze current water utility practices regarding main failures. A joint questionnaire was developed for two of WESTON's AWWARF projects to ensure better response and to minimize the burden on the water utilities. One part of the questionnaire addressed the requirements of this project while the other part was for AWWARF Project 457, "Guidance for Management of Distribution System Operation and Maintenance" (Deb et al. 2000).

The questionnaire was mailed to 70 utilities across North America. Some of these utilities had agreed to participate while the others were randomly selected from the WATER:\STATS database. Completed questionnaires were returned by 37 utilities. In Europe, 44 utilities were first contacted by LNEC to determine if they would participate. Twenty-nine of the utilities responded positively, and 28 ultimately returned the questionnaire. There also was collaboration with AWWARF Project 461 "Main Break Prediction, Prevention and Control", that included some of WESTON's questions in its questionnaire and vice versa. In this collaboration, a

WATER:\STATS random list of utilities was split into two to avoid any overlap in the utilities receiving the questionnaire. Both projects exchanged their results. As a result, partial information on 40 additional utilities was received through Project 461. A summary and comparison of the questionnaire results for North American and European utilities is presented in Table 2.1.

Of the 70 utilities in North America that were sent questionnaires, 12 served less than 100,000 people, 58 served more than 100,000 people. The response rate was higher for larger utilities. Five of the 12 smaller utilities responded (42%), while 32 of the 58 larger utilities responded (55%). The average daily water production (retail plus wholesale) of the utilities surveyed in North America was about 95 mgd, ranging from 2 mgd to 580 mgd. The length of their distribution systems varied from 100 miles to 7500 miles with an average length of 1628 miles of pipe. In Europe the utilities surveyed varied between 5 mgd and 535 mgd (20 mld to 2022 mld).

This survey found that between 1993 and 1997 the average break rate in North America was between 21 and 25/breaks/100 mi/yr. Similarly, Project 461's survey found that the average break rate was about 27/breaks/100 mi/yr. For the two surveys combined, the frequency distribution of the average break rates is presented in Figure 2.1. These charts show that about half of the North American utilities surveyed had 20 breaks/100 mi/yr or less. On the other hand, the average break rate in Europe is higher with about 50 breaks/100 km/yr (80 breaks/100 mi/yr). Furthermore, less than 20% of the European utilities have break rates as low as 20 breaks/100 mi/yr. One likely explanation is that most European utilities are in old urban areas, thus their infrastructure is far older than North American ones. A more in depth analysis by LNEC indicated that some of the high rates were either due to rapidly growing areas where extensive construction resulted in damage by other parties causing most of the main breaks (in southern Europe), or were due to poor construction and material quality.

Less than half of the North American utilities reported that they have a formal program in place for controlling main breaks, while 80% of the European utilities have such programs. While all European utilities have water main inventories, 11% of North American utilities did not have them. A water main inventory is critical to developing a comprehensive water main renewal program. Furthermore, 85% of European utilities have their main break data computerized, while 70% of North American utilities have them computerized. Figure 2.2 shows the relative age of main break data in North American databases.

Table 2.1
Questionnaire survey results

	Comments	
	North America	Europe
Utility size		
Water production		
• Average	95 mgd	305.5 mld (80.8 mgd)
• Range	2-580 mgd	20.3 - 2022 mld (5.4 – 535.2 mgd)
Total length of pipe		
• Average	1628 mi	4935 km
• Range	100-7500 mi	162 – 41834 km (100 – 26080 mi)
Main failures		
• Average	22 breaks/100 mi/yr	50 breaks/100 km/yr
• Range	1-102 breaks/100 mi/yr	<1 – 380 breaks/100 km/yr
• Formal programs for control of failure	46%	80%
• Main inventory	89%	100%
• Computerized main inventory of total	68%	95%
• Failure records	89%	90%
• Computerized failure records of total	70%	85%
General information		
• Date, address	96%	97%
• Temperature	22% air, 8% water, none soil	11% air, 11% water, 6% soil
• Time of detection and arrival	56%	45%
• Impact on surroundings		
• Services affected	37%	41%
• Blocks affected	22%	11%
• Hydrants affected	23%	11%
• Proximity to buried objects	25%	6%
Water main information		
• Material	96%	94%
• Location in street	93%	83%
• Diameter	93%	89%
• Depth	81%	44%
• Installation date	58%	61%
• Cathodic protection	38%	17%
• Joint type	58%	39%
• Type of repair	85%	67%
Pressure		
• Range	23%	18%
• Pump station status	19%	12%
Failures		
• Type of failure	95%	75%
• Probable cause	65%	50%
• Type of repair	88%	61%
• Exterior/interior condition of pipe	62%	44%
• Bell condition	25%	None

(continued)

Table 2.1 (continued)

	Comments	
	North America	Europe
Condition of valves		
• Required for isolation	50%	22%
• Condition	27%	17%
Reporting bedding conditions	42%	17%
Reporting of seismic/ geotechnical conditions		
• Soil description	38%	17%
• Geologic unit description	15%	11%
• Groundwater depth	12%	None
• Seismic hazard unit	8%	None
Collecting field samples		
• Pipe samples	32%	11%
• Soil samples	23%	None
Use of automated systems		
• Field portable computers	27%	10%
• GPS	37%	None
• GIS	67%	65%
• DBMS	85%	85%
Formal renewal program		
• Main replacement	85%	67%
• Main rehabilitation	42%	61%
Costs records		
• Direct labor	78%	56%
• Indirect labor	42%	50%
• Materials	78%	56%
• Equipment	72%	33%
• Surface repairs	69%	50%
• Damage	52%	44%

Note: Except for "utility size" and "main failures", the information provided is for utilities that have main break databases.

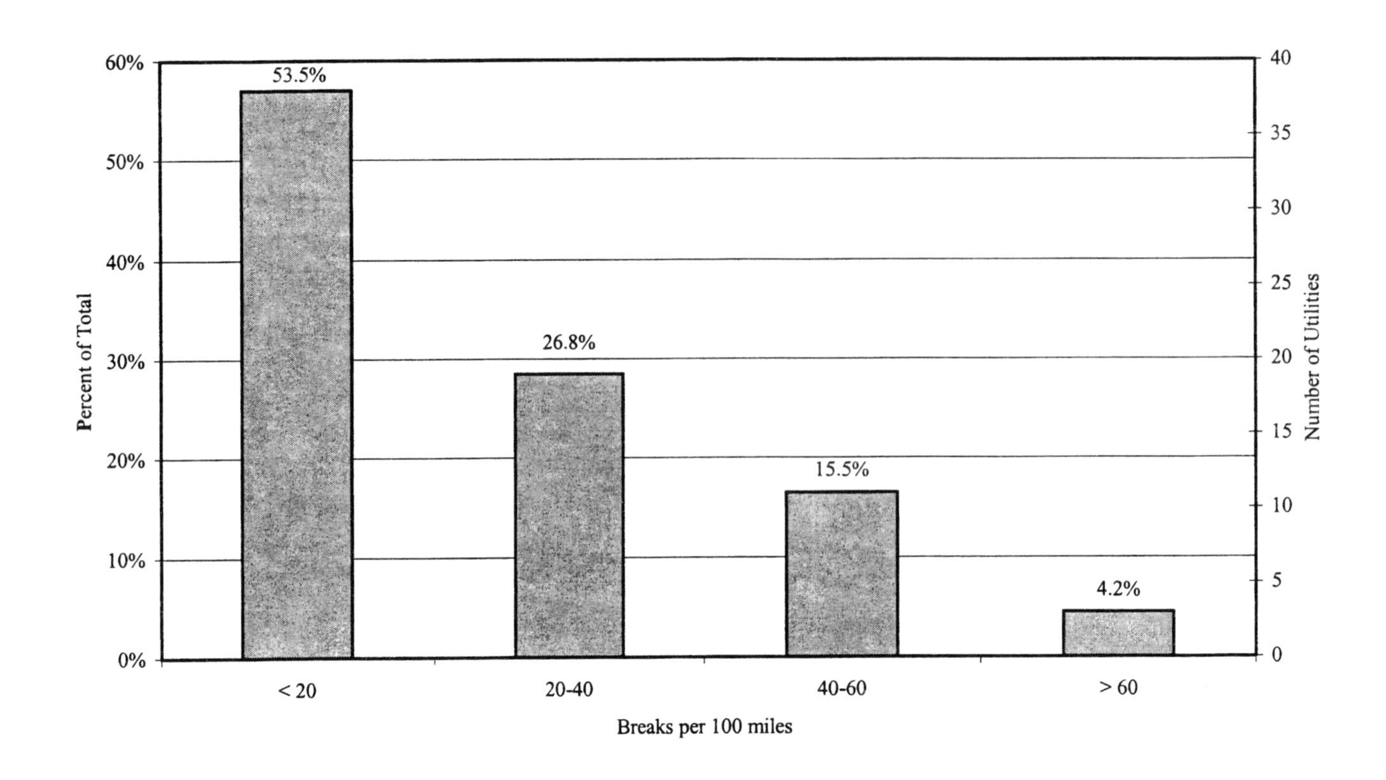

Figure 2.1 Frequency distribution of North American water utility failure rates

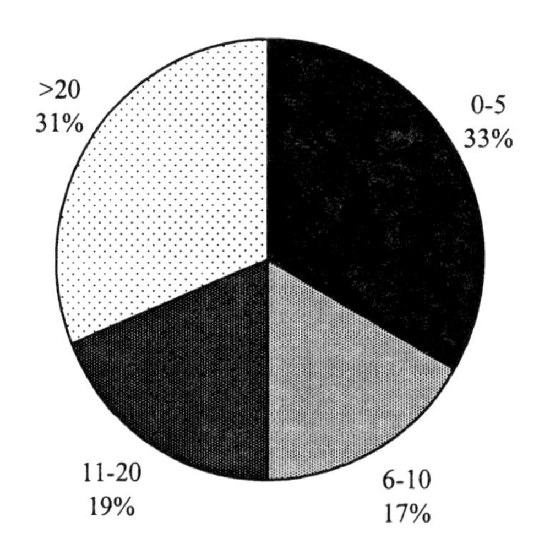

Figure 2.2 Number of years of water main failure data available in North American water utility databases

The collection of general information associated with main break events, such as date, address, time, temperature, impact of surroundings, etc., was similar in North American and European utilities. While date and address were recorded universally, very few utilities recorded any temperature data (air, water or soil) corresponding to the main failure. This is somewhat surprising since many utilities attribute main breaks to temperature changes. Information on proximity to buried objects such as gas lines, etc. was also scarce (25% for North America, 6% for Europe). Water main information such as material, size, and location on street was commonly recorded. Depth of pipe, however, is more commonly recorded by North American utilities (81%), but less so by European utilities (44%). Similarly, valve conditions, failure information, bedding condition were also recorded more commonly by North American utilities. Pressure information, seismic and geotechnical information such as soil conditions and depth to groundwater were not commonly tracked in either North America or Europe.

More North American utilities collected pipe samples for material testing (32% vs. 11%), but few American (23%) and no European utilities collected soil samples. Practices in the use of automated systems were similar in geographic information system (GIS) and database management system (DBMS) applications, but North American utilities were more active in the application of portable computers for field use and global positioning system (GPS). Tracking the costs of water main failures was practiced by about one-half to three-quarters of the utilities, with it being more common in North America.

UTILITY MAIN BREAK DATABASES

In order to gather additional information on main breaks and related data, the survey questioned the water utilities whether they had electronic main break databases. Most water utilities maintain a record of water main breaks in one form or another. These records may simply be paper forms filled out by office personnel and field crews in responding to main breaks. However, a more common approach is to create and maintain computerized databases of main breaks. These databases are often maintained and updated by the distribution department for their own use. Typically, older main breaks, i.e. those that occurred prior to the creation of the database, are not entered into the database even though the paper records may still exist. In some cases, the main break data is compiled as part of an overall work management system (WMS). The WMS may include information from the initial call to customer service, to the generation of a work order directing a crew to respond, to the recording of crew times for payroll purposes. A sophisticated system may include links to a GIS to provide spatial capabilities.

The availability of a main break database is critical for evaluating the causes of main breaks and for planning future distribution system renewal programs. The contents of the database will vary from utility to utility, but certain key fields of data are necessary to successfully utilize the database for planning purposes.

Based on the responses received from the questionnaire, water main break databases were obtained from PWD, Ft. Worth Water Department (FWWD), St. Louis County Water Company (SLCWC), Philadelphia Suburban Water Company (PSWC), Regional Municipality of Ottawa-Carleton (RMOC), and United Water New Jersey (UWNJ). These databases provide actual data for use in developing and calibrating the predictive main break models. Each of the utilities uses different database management software, and each provided the data in different formats. WESTON imported each utility's data into an MS Access database for use on this project.

PWD maintains a mainframe database of main breaks. Electronic data are available dating back to 1964, and the database contains approximately 29,000 records. FWWD provided approximately 60,000 records dating to 1971. SLCWC has an electronic database dating to 1983, and also has paper records of main breaks dating to the early 1960's. The PSWC database has approximately 18,000 records dating back to 1960. RMOC's database includes approximately 6,500 records of water main breaks from 1972 through the present. UWNJ maintains separate databases for water main leaks and breaks. The leak database contains approximately 3,000 records dating to 1968, while the break database contains approximately 4,000 records from 1963 through the present. Both databases share the same general format and content.

The following general categories of data are typical of data maintained in a main break database:

- Background data – date, time, address, etc.
- Field observations – specific location, condition of pipe, cause of break, etc.
- Pipe history – installation date, work order or other pipe identifier, etc.
- Cost data – labor, material, and equipment, etc.
- Testing results – soil testing and pipe testing

Table 2.2 summarizes the general data available in the main break databases provided by the six utilities.

Review of these databases provides a view of current water utility practices in managing water main break data. It is interesting to note that of the six utilities only PWD has historically incorporated materials testing in main break analysis. This was part of a special study and is no longer routinely performed. Most of the utilities track time and materials as part of the main break database, providing an opportunity to directly track the "costs" of main breaks. In most cases, detailed cost information is tracked separately from the main break information, making it difficult to determine the costs of a specific main break incident.

These databases provided raw data for use in developing and calibrating the predictive models discussed in Chapter 4. They also served as a check that the data requirements for modeling were practical and within the ability of water utilities to produce.

PIPE RENEWAL PRIORITIZATION METHODOLOGIES

A review of the literature identified several basic approaches for prioritizing pipes for replacement or rehabilitation within a distribution system. These may be grouped into the following four categories:

- Deterioration point assignment methods
- Break-even analyses
- Failure probability and regression methods
- Mechanistic models

Table 2.2
Summary of main break databases provided

Utility	Background data	Field observations	Pipe history	Cost data	Testing results
PWD (Philadelphia Water Department)	Date, location (address, plate #, census tract, etc.)	Detailed location (distance from curb, etc.), pipe material, pipe diameter, break type, joint material, pipe condition (corrosion, tuberculation, wall thickness, etc.), soil condition, frost depth, pavement condition, repair type	Installation year, replacement contract #	Repair time, main out of service time, property damage, number of services affected	Sample ID (link to separate test result database)
FWWD (Ft. Worth Water Department)	Break ID, date, address	Pipe material, pipe diameter, break type, repair type	None	None	None
SLCWC (St. Louis County Water Company)	Ticket #, date, address, map information (grid)	Break type, indication of corrosion, pipe diameter, pipe material, joint type, ground surface (paved, grass, etc.), detailed location (distance from intersection, etc.), depth of pipe, number of anodes installed, frost depth, other utilities, chlorine checked	Work order, installation year, length installed	Repair time, material used	None
PSWC (Philadelphia Suburban Water Company)	Date, location (division, municipality)	Detailed location, pipe material, pipe diameter, leak type, repair type, depth, apparent pipe condition	Extension number	Repair time, capital cost, damage caused, shut down duration, customers affected, number of valves closed	None
RMOC (Regional Municipality of Ottawa-Carleton)	Date, location (district, map #), township, pressure zone	Pipe material, pipe diameter, break type, pipe depth, frost depth, bedding, pipe condition, soil type	Pipe ID, length, installation year, C-factor	None	None
UWNJ (United Water New Jersey)	Date, location (municipality, map #, etc.), pressure zone	Detailed location, pipe material, pipe diameter, type of break, cause of break, pavement type, soil type, depth, traffic conditions	Extension number, installation year	Time required, materials used	None

Deterioration Point Assignment Method

In the deterioration point assignment (DPA) method, a set of factors associated with pipe failures is identified. These factors may include age of pipe, pipe material, pipe size, type of soil, location, water pressure, discoloration, and number of previous breaks. The numerical values for these factors are grouped into several class intervals. For each class interval a failure score is assigned. For any pipe a total failure score is obtained by summing the class interval failure scores for that pipe. If the total failure score exceeds a threshold value then the pipe is considered a candidate for replacement or rehabilitation. This approach cannot discriminate between competing pipes when funding is limited. That is, pipes receiving the same score cannot be further prioritized. Also, this assessment lacks the predictive power that is crucial for future course of actions.

An example of this type of system was described in Deb et al. (1995). The Louisville Water Company's (LWC) Pipe Evaluation Model (PEM) includes a detailed scoring system that assigns points based upon 23 parameters. The 23 parameters encompass four broad categories, namely geographical, service quality, hydraulic and maintenance.

Break-Even Analysis

The break even analysis is a cost based method and considers repair costs and replacement cost simultaneously. This method must be augmented with a predictive technique for pipe breaks. The predicted break occurrence times are utilized along with the replacement cost. The present value cost of replacing the pipe decreases over time. The present value of cumulative repair costs increases over that same time period. At any time, the total cost associated with the pipe is the sum of the present values of replacement and cumulative repair. The optimum economic time to replace the pipe is when the total present worth cost is at a minimum. Stacha (1978), Shamir and Howard (1979), Walski (1987) and Walski and Pellicia (1982) have addressed the break-even analysis in more detail.

In determining whether to replace or repair a failed pipe, Stacha (1978) compared the annual cost of replacing a pipe to the annual cost of repairing. A methodology for using historical record of failures and repairs to project the accumulated cost of repair within a specific time frame was presented. The accumulated cost was then compared with the replacement cost to make a repair decision. Stacha advised that use of the cost difference alone was inadequate. Other parameters such as water quality and flow capacity also needed to be taken into consideration.

Male et al. (1990) described an effective replacement policy in use by New York City. The authors showed that at that time, the policy of replacing all pipes with two or more breaks was the most cost effective system-wide. The analysis involved the consideration of five alternatives: (1) replace after one or more breaks, (2) replace after two or more breaks, (3) replace after three or more breaks, (4) replace after four or more breaks, and (5) do nothing. Alternative 2 also turned out to be the most proactive policy. The authors also showed that some cost improvements could be made if alternative replacement policies were used in some of the boroughs. Male et al. (1990) also indicated that the discount rate used in the calculation affected which alternative was selected. For example, higher discount rates lead to a less proactive policy and vice versa.

Kleiner et al. (1998a, 1998b) considered repair costs in terms of improved flow capacity of a pipe achieved by relining. A reduced C-value would increase the head loss and reduce pressure. Kleiner et al. incorporated mass balance, energy, and head-loss with time varying C-value and minimum pressure constraints in a rolling time horizon to find the optimal replacement times as relining costs become non-optimal. The objective was to minimize total costs of relining and associated replacement cycles in an infinite time horizon.

Su et al. (1987) and Wagner et al. (1988) addressed an alternative form of a reliability constraint based on the probability of satisfying nodal demands and pressure head requirements under various network failure configurations. Duan et al. (1990) considered the optimization and reliability of pumping systems in a network. Similar studies were also presented by Lansey and Mays (1989), Mays (1989), and Park and Liebman (1993). Although these models did not directly address the causal mechanisms of failures in pipes, they provide a framework for further economic analysis. Issues and methods related to pipe network optimization are fully covered by Loganathan et al. (1990, 1995) and Sherali et al. (1998).

In 1999, SLCWC conducted an economic evaluation of water mains in its distribution system. Grablutz and Hanneken (2000) described an economic model that considered the total present worth cost of a pipe to be the sum of cumulative projected future repair costs plus replacement cost. The optimum economic time to replace a pipe was when the total cost was at a minimum. The authors also noted that non-economic factors also would certainly be considered in the final replacement decision.

Failure Probability and Regression Methods

The failure probability methods are related to the DPA method in that they build on the same deterioration factors but bring in a predictive capability by assessing the probability of a pipe's survival. However, only a few of the failure probability models are well detailed. Comprehensive reviews of such models are given in Andreou et al. (1987) and O'Day et al. (1986). The two lumped parameter models of Shamir and Howard (1979) provided a rudimentary estimate for the number of breaks at a chosen time. In an attempt to account for the relative impacts of various exogenous agents, Clark et al. (1982) proposed certain multiple regression equations for the number of years from installation to the first repair. Another equation was also proposed for the number of repairs over a time period measured from the time of the first break. These equations had coefficients of determination (R^2) of 0.23 and 0.47, respectively. Thus, while Clark et al.'s procedure was a significant improvement in predicting pipe breaks, it did raise some concerns because of the low values of the coefficients of determination. Following Clark et al.'s study, Andreou et al. (1987) suggested another approach to estimate the survival probability of an individual pipe with the aid of Cox's regression model. In contrast to Clark et al.'s expected times of failures, Andreou et al. provided the probabilities of failure. Clark et al. made the following observations: only a subset of pipes have recurrent repairs; the time to first repair is relatively long (typically about fifteen years); the time between repairs become shorter as pipes get older; large diameter pipes tend to have fewer problems; and industrial development, in general, results in more repairs. Goulter and Kazemi (1988, 1989) observed both spatial and temporal clustering of pipe failures for the city of Winnipeg. A non-homogeneous Poisson distribution model was proposed to predict the probability of subsequent breaks in a pipe given that the first break had already occurred. Mavin (1996) provided a review of the failure models in the literature. Mavin also pointed out the need to filter the data before constructing a failure model. He suggested to not including breaks that occurred within three years from installation and six months from a previous break, because these types of failure were likely to be associated with construction faults and not with a structural failure of the pipe. Based on the filtered data, a set of regression equations was constructed for number of failures over a time period and time interval between breaks.

Deb et al. (1998) discussed a probabilistic model called KANEW to estimate miles of pipes to be replaced on annual basis. The primary object of KANEW was to provide water utilities with a tool to develop their long- range pipe renewal strategies. Based on the historical inventory of water mains and the estimated life-span data, KANEW predicted miles of different categories of pipes to be rehabilitated and replaced on an annual basis. It was not, however, intended to provide location specific rehabilitation and replacement information. The model used the actual water main inventory, with the pipes categorized according to such basic characteristics as age, material, diameter, soil corrosivity, etc. For each category 100th, 50th and 10th percentile ages were obtained either by expert opinion or by analysis. These percentiles were utilized to obtain the three parameters of the Herz probability density function from which the survival probabilities were obtained. These survival probabilities were used to obtain the expected survivors or its complement of non-survivors per year, which were to be renewed.

Roy F. Weston Inc. and TerraStat Consulting Group (1996) developed the Pipe Evaluation System (PIPES) model for use by the Seattle Public Utilities to evaluate the rehabilitation needs of pipes in the system. The PIPES model consisted of three sub models: deterioration, vulnerability, and criticality. Input data to these models were provided by a GIS database to correlate pipes with layered spatial data. The deterioration model used statistical analysis to relate pipe break history data to pipe properties (type, size, etc.). This analysis relied on the Cox regression proportional hazards and the Wiebul distribution. The deterioration model provided the probability of failure of a specific pipe within the next ten years. The vulnerability model further refined the results obtained from the deterioration model. Typically, the vulnerability model provided an answer to a question like: "Which pipes are most likely to fail in the future?" The criticality model identified critical pipes based on their failure costs and the kinds of facilities they served. Thus, this model combined aspects of failure probability and cost models.

Shamir and Howard (1979) used regression analysis on pipe break data to develop an exponential function that predicted the number of pipe breaks for any given year. The costs associated with pipe repairs over a chosen number of years were compared with the replacement cost and the optimal year of pipe replacement was determined.

Walski and Pellicia (1982) proposed a model based on the use of pipe break history to project the pipe failure rates. Their model closely resembled the one by Shamir and Howard (1979) but had some modifications. Walski and Pellicia provided a correction factor to be applied to pipes based on the size and type. The factor also accounted for the effect of breaks due to cold temperatures. This temperature correction factor was related to the average temperature of the coldest month. Walski and Pellicia warned that due to the difficulty in predicting the severity of a winter until after the fact, the use of the temperature correction factor in predicting future breaks could be erroneous. Walski (1987) improved on the previous research by introducing a cost model that accounted for the lost water due to leakage and broken valves.

Karaa et al. (1987) described an appropriate technique for time phased replacement of failure-prone pipes. In this procedure pipes that were to be rehabilitated, replaced, or to be constructed were grouped into so-called "bundles", based upon similarities in characteristics and criticality measures. A failure model might be used to delineate such bundles, based upon their role in determining the reliability of the system. Furthermore, the optimal number of bundles to be replaced was determined through the use of a linear program with annual budget constraints. Sensitivity analyses could be performed to assess the variations in proportions of pipes to be replaced/rehabilitated each year against budget changes. Such analyses also pointed to the budgetary requirements for various anticipated levels of system upgrades. This information was crucial for proper planning, and assisted in understanding the gravity of the problem. For example, this type of analysis could show that to ensure a reliable network by the year 2020, a water utility should rehabilitate or replace 11 miles of pipe annually. In the same vein, PPK Consultants (1993) provided a comprehensive assessment procedure considering water distribution system as a critical resource.

Lane and Buehring (1978) discussed LADWP's use of a DBMS to formulate a "sound long-range" replacement program. The computer system used information on pipes (properties and service history); information on their surroundings (soil and water properties); and potential for liability to prioritize pipes for rehabilitation. The LADWP DBMS was used to identify pipes with high probabilities of failure and required engineering judgement to select a specific pipe for replacement.

Another failure probability study is the development of UtilNets by the Computer Technology Institute (CTI) and funded by the European Union (CTI 1997). UtilNets was a decision-support system (DSS) for rehabilitation planning and optimization of water distribution networks, and uses expert system (ES), DBMS, and GIS tools. The DSS performed reliability-based life predictions of pipes and determined the consequences of maintenance and neglect over time in order to optimize rehabilitation policy. The prototype of UtilNets was implemented for gray cast iron water pipes, but could be extended to other pipe materials. Since complete information about the state of the pipe network was generally not available, UtilNets was designed to yield reliable forecasts even when data were incomplete. It included the following (CTI 1997):

- "Probabilistic models that give a measurement of the likelihood of structural and hydraulic failure of pipe segments over the next several years
- Assessment of the effects that pipe condition can exert on water quality
- Assessment of both the quantifiable and qualitative consequences of various rehabilitation options and neglect over time
- Selection of the optimal rehabilitation policy for each failed pipe segment
- An aggregate structural, hydraulic, water quality and service profile of the network together with an assessment of the required rehabilitation expenditures
- An assessment of network reliability in terms of demand point connectivity and flow adequacy"

Mechanistic Models

Several mechanistic models or approaches have been used to model corrosion (Romanoff 1957, Rossum 1969, Kumar et al. (1984, 1986, 1987), and Basalo 1992). For modeling the change in pit depth with time, soil environment and age, Rossum (1969) developed a set of equations. Rossum's equation for the pit depth had the form:

$$p = f(\text{soil parameters}) * \text{time} * [(10\text{-pH})/\text{soil resistivity}]^{N} \qquad (2.1)$$

where p = pit depth

N = parameter

His equations are partly based on the extensive data collection effort by the National Bureau of Standards (NBS) (Romanoff 1957). The NBS buried 36,500 specimens representing 333 varieties of materials in 47 soils starting in 1922. An analysis by NBS led to an equation of the form:

$$p = k(T)^n \qquad (2.2)$$

where p = depth of the deepest pit at time, T

k, n = parameters

The values of the parameters k and n were provided for the 47 different soil groups with the fit being considered poor for only four. Later, Rossum (1969) took advantage of these results in developing his equations.

These mechanistic models can be categorized by the major stress considered in the model. Various models have been developed for temperature-induced stresses (U.S. Pipe and Foundry Co. 1962, Wedge 1990, and Habibian 1994), frost load (Cohen and Fielding 1979, Fielding and Cohen 1988, Rajani and Zahn 1996) and other failure processes using the underlying physical principles.

PWD (1985a) provided a detailed account of the various structural failure modes for a water main. Rossum (1969), Kumar et al. (1984) and Ahammed and Melchers (1994) suggested approaches for modeling corrosion.

Kumar et al. (1984) provided a methodology for assessing corrosion growth in terms of a Corrosion Status Index (CSI) over time. The CSI depends on pipe coating, liquid carried, buried depth, soil resistivity, soil chlorides, soil sulfides, soil pH, soil moisture, pipe material, cathodic protection, and pipe diameter. The CSI is given by:

$$CSI = 100 - 100[P_{av}/t] \qquad (2.3)$$

where CSI = Corrosion Status Index

P_{av} = average pit depth of a 1 m section of pipe

t = wall thickness of pipe

The pit depth, P_{av} and CSI enable one to predict the loss of useful pipe thickness over time, which in turn is related to the residual strength of a pipe. If the loss in thickness is denoted as d, the design thickness as t, and failure thickness as t_f, then $P[t\text{-}d<t_f]$ indicates the probability of failure for the pipe of interest. By relating the current thickness (t-d) to its residual strength, S_y, from laboratory tests a comparison can be made against the anticipated applied stress, S_{act}. The risk can also be evaluated as the probability that the strength, S_y, is less than the applied stress, S_{act}, denoted by $P[S_y \leq S_{act}]$. Another measure is the safety factor defined as the ratio of residual strength, S_y, to the applied stress, S_{act}.

Besides external corrosion, water mains are also prone to deteriorative mechanisms occurring internally. Through experiments Millette and Mavinic (1988) showed that the internal pipe deterioration through corrosion is dependent upon certain water quality parameters. These include pH, water velocity (or pressure), hardness, and dissolved oxygen content. Millette and Mavinic reported the following findings: cast iron corroded twice as fast in a pressurized system as opposed to in a gravity system and iron levels found in tap water exceeded levels found in raw water indicating the presence of corrosion and iron uptake by the water in the distribution system.

Wedge (1990) showed that the excess pressure developed in a piping system could amount to as much as 200 psi for a 10° F change in water temperature. Wedge's methodology basically converted the strain due to thermal expansion into to a corresponding pressure. Pipe break data analysis of the Washington Suburban Sanitary Commission (WSSC), Md., water distribution system showed a trend in increased pipe breakage rate due to temperature changes (Habibian 1994). Specifically, the data showed that temperature drop was a significant factor in increased pipe break rate. Each time the temperature reached a new low; a surge was noted in the number of breaks. Such a trend indicated that the temperature change alone was inadequate to correlate increased number of pipe breaks. Possibly, the actual initial and final temperatures had to be considered.

Temperature also affects the pipe in the form of increased loads that results from frost heaving of the soil. Monie and Clark (1974) were among the first to show experimentally the increased load exerted on pipes by frost heaving. In an experiment conducted in Portland, Maine, Monie and Clark found that the load on the buried pipe doubled due to frost heave. Also, frost conditions seemed to transmit live loads to the pipes from farther distances from the buried pipe. Though the authors attributed increased number of breaks in pipes to frost loads, they also speculated that the cold water had the potential for increased stresses in the pipe thus leading to more failures.

Cohen and Fielding (1979) provided a simplified formula for the determination of the frost depth in a soil as a function of the freezing index. Fielding and Cohen (1988) further developed a modified Boussinesq equation relating the expected frost load with the frost depth. The results obtained from the modified Boussinesq equation compared very well with field measurements.

In frost load assessment Rajani and Zahn (1996) suggested computing frost heave using two paths due to a) freezing of the in situ pore water, and b) water arriving at the freezing front from elsewhere. With the aid of a vertical one-dimensional force equilibrium analysis, the frost heave was converted into incremental frost pressure at the freezing front.

While these mechanistic models help to understand the failure processes, the predictive capability must be considered either through a correlation analysis or through a probabilistic analysis by considering the parameters/variables to be random. As described in Chapter 4, in this study the mechanistic methods provide strength estimates as a function of failure causing factors. The mechanistic methods are incorporated within a probabilistic scheme for assessing the stage of deterioration of an underground pipe. The stage of deterioration has to be inferred from the environmental factors and repair history of a pipe. The background deterioration is attributed to corrosion. However, this failure rate can be accelerated by unintended traffic load, frost load, and temperature effects. The accelerated failure rate is accommodated by shifting between curves or by raising the failure thickness. The net result is that it is possible to obtain the residual strength for a pipe at a given time. It is precisely the information needed for making replacement and rehabilitation decisions.

GAS UTILITY PRACTICES

Gas and water utilities face similar challenges in conveying a product to the customers of the system. The obvious difference is the material that is conveyed through the pipe. Due to its explosive nature, gas pipes are managed and monitored more closely to avoid failures and their resulting consequences. Similarly, the operation and maintenance of gas distribution systems is more regulated than water distribution systems. High priority is given to public safety and customer service in gas distribution system where federal regulations are in place for all aspects of operation and maintenance. Some of the major operation and maintenance features of a gas distribution system that may be applicable to water distribution systems include (Deb et al. 1995):

- Unaccounted - for product (comparable to unaccounted - for water) in the gas industry is closely monitored and is generally less than 5 percent.
- Regulations require gas utilities to conduct leakage surveys on a regular basis. The frequency of the surveys depends on the service district and consequences of leaks on public safety. Annual surveys are required for downtown and high value districts.
- Every time a gas pipe is exposed, its condition must be assessed and recorded. As a result, a gas utility typically has extensive data on the condition of pipe in its system.
- All gas utilities are required by regulation to submit operations and maintenance (O&M) plans.
- Gas utilities classify leaks on the basis of consequence and location. A leak that poses high risk for public safety is classified as a Grade 1 leak and must be repaired immediately. A leak that does not represent a safety risk is a Grade 2 leak that must be monitored regularly until a repair is completed.

Unlike water utilities, the documentation requirements for a leak incident in a gas utility are well established. A leak report for a gas utility would as a minimum include the following:

- Work order #
- Street address

- Plate #
- Condition of pipe
  - Excellent
  - Good
  - Fair
  - Poor (immediate notification is required if pit depth is about 70% of wall thickness and should be clamped.)
  - Graphitization
- Repair description
- Number and size of openings
- Tax district
- Town code
- Leak grade
- Pipe size
- Pipe material
- Cause of leak
- Type of coating
- Condition of coating
- Anodes installed
- Time
- Date
- Type of pavement removed

A leak record card process flow chart as followed by a southern New Jersey gas company is shown in Figure 2.3. After completion of a leak repair, information from the leak record cards is used to prepare leak maps. As a minimum leak maps contain the following information:

- Repaired leaks
- Documentation of leaks
- Information on main replacement and improvements

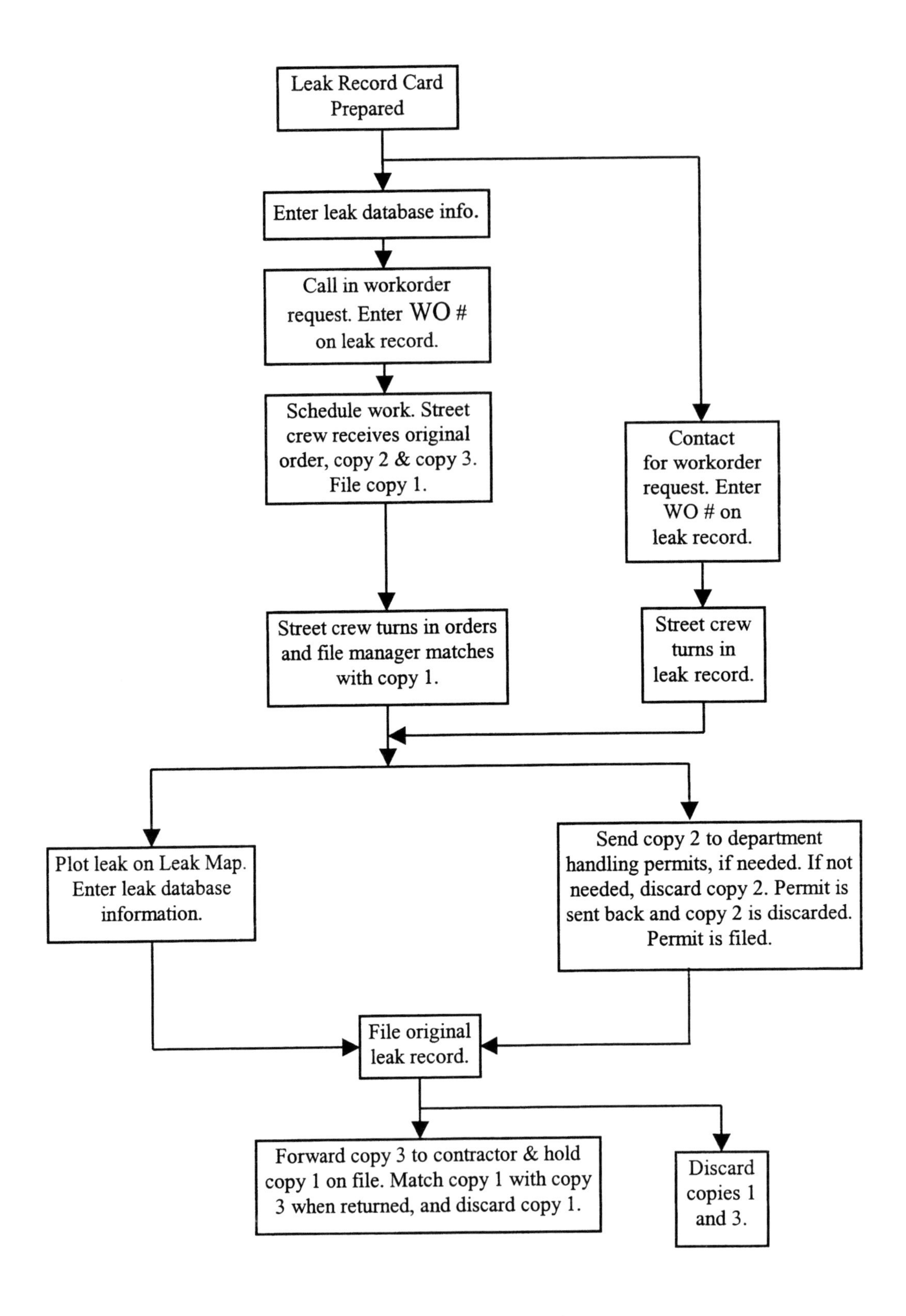

Figure 2.3 Sample leak record card process flow chart for a gas company

The leak maps are kept up to date as part of the normal leak record documentation. The leak maps are also used in conjunction with main replacement evaluations that are used to prioritize infrastructure renewal.

Gas utilities often apply the concepts of risk management in optimizing maintenance activities and in developing infrastructure renewal programs. Risk management is used in a variety of fields to achieve an acceptable level of operational safety at an affordable price. In a manufacturing setting, the processes, manufactured components, and other factors that contribute to malfunctions are controllable in a way that detailed cost/benefit analyses can be performed to optimize the process and minimize risk. For the gas industry, risk management involves balancing the consequences of a gas main or pipeline failure against the costs of inspecting, maintaining, rehabilitating, or replacing the pipe(s) in question. Research is being conducted by the Gas Research Institute (GRI) to develop formalized procedures for assigning probabilities of risk. Until the formalized procedures are available, the industry has been using a "ranking" system based on "relative" risk to prioritize maintenance.

Kiefner and Morris (1997) describe a Risk Management Tool (RMT) that consists of an algorithm which calculates a total risk number (R_{tot}). R_{tot} is the product of P_{tot} and C_{tot} where:

P_{tot} = sum of the probabilities of failure from all postulated causes, and

C_{tot} = consequence of failure comprised of consequences to life and property, to throughput, and to the environment

The consequences of a gas main or pipeline failure can be more severe than a water main break. However, many of the concepts used in estimating the probability of gas main failure are applicable to pipes in a water distribution system. The following components are considered in estimating the probability of failure for a gas main:

1. External corrosion – The probability of failure due to external corrosion is a function of wall thickness, corrosion rate factor, stress (pressure) level, history of previous incidents, temperature, stray current, coating type, coating condition, microbiological influence in surrounding soils, and mitigating influence of inspections or testing. For

external corrosion and many of the other components that follow, the probability of a failure is assumed to be reduced as a result of inspections or testing of the pipe.

2. Internal corrosion – Many of the same factors that influence the probability of external corrosion causing a failure apply to internal corrosion. These include wall thickness, stress (pressure) level, history of previous incidents, and mitigating influence of inspections and testing. In addition, gas composition, biological activity within the product, inhibitor effectiveness, and seam orientation (certain weld types promote internal corrosion) must also be accounted for in considering internal corrosion.
3. 3rd party damage – The probability of failure due to 3rd party damage is dependent on pipe geometry (wall thickness and diameter), pipe material strength, class location, depth of burial, history of prior encroachments, and preventative response factors. Class location refers to the land use and population density in the vicinity of the gas main, and is defined uniformly across the gas industry. Preventative response refers to the effectiveness of "one call" systems, patrolling of the right of way, and permanently marking or identifying right of ways.
4. Stress-corrosion cracking (SCC) – SCC is characterized as an environmentally-stimulated, stress-driven form of longitudinally-oriented cracking. It occurs only in pipes subjected to very high hoop stresses, and is very temperature dependent. Factors considered in estimating the probability of failure due to SCC include stress (pressure) level, mitigating influence of testing, distance from compressor station (which impacts pressure and temperature), stress corrosion indicator, coating type, coating condition, history of prior SCC incidents, and temperature.
5. Material factors – The probability of failure due to material factors (i.e., pipe manufacturing defects) is a function of the ratio of initial hydrostatic testing pressure to operating pressure, seam type, age of material, girth weld quality, history of prior material incidents, and mitigating influence of retesting.
6. Construction factors – Construction defects might include weld defects, dents, gouges, and poor bedding or installation. Most construction factors are accounted for in other probability equations (for example, 3rd party and material factors), and thus

there is no formal accounting for construction factors unless prior history has shown a problem associated with a particular construction technique or company.

7. Outside sources – This factor accounts for the probability of failure due to natural forces in locations such as water crossings, landslide or subsidence areas. Specific factors include amount of crossing with shallow burial, amount of crossing with extremely shallow burial, amount of span subject to undermining, scouring potential, and amount of unsupported span.
8. Electrical disturbances – Some pipelines may be at risk from various kinds of electrical disturbances such as lightning strikes, electrical induction in areas with high voltage power lines, or mining areas where ground return might be used for dc equipment. Like the construction factors situation, some of the probability of failure due to electrical disturbance is accounted for in other probability equations, and is not considered separately.

All of these factors used in the gas industry are also applicable to pipes in a water distribution system. As a result, a similar approach of assigning probability of failure to water pipes could be developed as a first step in planning future maintenance activities and ultimately for prioritizing water main renewal programs.

The need to protect public safety has resulted in a greater degree of regulation in the gas industry. The types of data being collected by gas companies related to the condition of pipes are not significantly different than the data generally being collected by water utilities. The major difference between the two industries is that a typical gas utility has consistently collected a uniform and comprehensive set of data related to the condition of its piping infrastructure. This allows them to anticipate failures and plan future renewal programs more effectively.

SUMMARY

This review of current information identified current practices related to water main break data collection. It also highlighted uses of these main break data, primarily in terms of developing water main renewal programs.

The WATER:\STATS database does not contain detailed data on water main breaks. The database does show that almost half of the distribution system pipe in the US and Canada is cast iron. It also provides the number of water main breaks that occurred annually. In 1995 the 702 utilities that responded to the question experienced 90,952 main breaks, or approximately 23 breaks/100 mi/yr.

The utility questionnaire gathered data on current water main break data collection practices in North America and Europe. The survey found that break rates are generally higher in Europe (50 breaks/100 mi/yr) than in North America (22 breaks/100 mi/yr). The higher break rates may explain why 80% of European utilities reported that they have formal programs in place to control water main breaks, compared to less than half of North American utilities. About 11% of the North American utilities surveyed do not have water main inventories. This hinders the development of water main renewal programs in these systems since data on existing water mains is not easily accessible. On the other hand, 85% of European and 70% of North American utilities reported that they had water main break data in some form of computer database. This is an important first step in putting the data to use in the development of water main renewal programs.

The questionnaire also examined the types of water main break data that are collected. Basic information such as date, location, and type of break is recorded almost universally by those utilities that collect water main break data. Few utilities in North America or Europe record air, water, or soil temperature. This is somewhat surprising because temperature is often cited as one of the major factors influencing main breaks.

Actual water main break databases were collected from six utilities. These databases showed a wide range of detail, and are typical of water main break databases in the industry. A review of the databases showed that much of the detailed data needed to conduct a systematic analysis of main break trends is often not collected by utilities.

The literature review identified numerous methodologies for prioritizing water main renewal programs, many of which relied to some extent on the availability of water main break data. The DPA systems use a scoring methodology to rank pipes for renewal. The number of previous main breaks is typically an important scoring factor. Economic models compare the costs of renewal to ongoing repair costs for a pipe. Water main break data must be available to apply such models, including detailed cost information. Failure probability and regression models require very detailed water main break data in order to effectively analyze and identify the underlying causes of main breaks. A large data set is also valuable in order to improve the validity of the analyses. Mechanistic models make the most use of water main break data, and also require the most detailed data. These models attempt to predict water main breaks by considering the characteristics of the pipe and its environment, and simulating the loads the pipe experiences. Thus, the water main break data must include related information such as soil characteristics, temperatures, pressures, etc.

Finally, gas utility practices relative to pipe failures were examined. The more regulated environment of the gas industry has led to a more standardized approach to collecting information when a pipe fails. The types of information collected when a gas main fails are not substantially different from what a water utility might collect. However, because these standardized rules have been in place for some time, gas utilities typically have a larger database at their disposal. This facilitates the analysis of the data and provides a more comprehensive picture of the condition of the distribution system pipe.

CHAPTER 3

WATER MAIN BREAK DATA COLLECTION

Water main breaks are burdens for all water utilities. They represent a drain on limited maintenance budgets, create water quality and pressure problems, inconvenience customers, and can cause considerable property damage. However, they also represent opportunities to examine distribution system pipes that have been out-of-sight for many years. These examinations can provide insights into the condition of the distribution system, and the information gained can help a utility effectively plan infrastructure renewal programs. In order to take advantage of these opportunities, a utility must recognize the types of data that should be collected and maintained every time a water main fails. This chapter describes the basic types of data that can be obtained while responding to a main break, discusses how to best collect and maintain that data, and presents the findings from field work performed to test these procedures.

DATA OPPORTUNITIES

Various types of useful data can be collected every time a water main fails. The first type of data is field data and includes the basic information about the failed pipe and its surroundings. This includes such basic physical information as diameter, material, depth, type and cause of break, water temperature, soil conditions, etc. The second basic type of data is referred to as office data. This includes background information about the pipe that cannot be observed in the field. This type of data includes installation date and previous break history for the pipe. A third type is laboratory and field test data of pipes and soils. It may not be practical or affordable to test all pipes that fail and the soil surrounding them. But representative sampling of pipe and soils can yield valuable information for planning renewal programs. The final data type involves the costs of water main breaks. Cost data includes both direct and indirect costs associated with repairing the water main. Direct costs such as labor and material costs are commonly tracked for all maintenance activities. Other costs associated with water main breaks such as damage claims and overhead time associated with data management should also be captured. Ideally, these data would also include indirect costs such as customer inconvenience, traffic disruption, etc. These intangible costs are difficult to quantify, but should be considered in the decision making process when developing a water main renewal program. All of these types of data are necessary for the utility to take full advantage of main break data in planning for the future of the system.

Each utility should design its own specific data collection program. The design should consider the goal of the data collection program (i.e., What is the data going to be used for?), the resources available (i.e., Will field crews have the time to fill out forms? Are office staff available to research and record background information?), and existing data collection practices and available databases within the utility. This last consideration is important because a great deal of information needed to effectively analyze the causes and impacts of water main breaks is often already being collected within a water utility. The challenge for the utility is to identify the location of the data, access the data, and link the data to specific water main break events.

It is important to define what is and what is not considered a water main break relative to this study. In many utilities one group is responsible for investigating and repairing "leak" and "break" incidents in the distribution system. Therefore, the collection and management of information related to these incidents are often performed concurrently. For this study, a water main break was defined as a *structural failure of the pipe*. Pipe failures at a joint were considered water main breaks under this definition, but leakage at a joint did not qualify as a main break. Breaks or leaks in service lines or fire hydrant laterals also were not included. However, many of the basic data collection and management concepts also apply to distribution system leaks service line and fire hydrant lateral breaks or leaks.

Before discussing the specific data that should be collected, it is illustrative to consider the typical sequence of events that are associated with the identification, investigation, repair, and follow-up of a water main break. Each step in this process generates information or data that should be recorded and maintained for later use. The availability of information, and the practicality of collecting it, must be considered by each utility in deciding which part of its organization is best suited for gathering the data.

A detailed discussion of the steps involved in responding to a water main break, including the additional steps that may be required in the event of a large main break, is provided in Appendix A. The specific activities associated with each step are likely to differ among utilities. The following basic steps are typical of water utility's response to a main break situation:

- Identification of problem – A call from a customer of the system or an observation by a utility employee are typical examples of how a distribution system problem is first noted. Investigation of the problem will be initiated immediately or scheduled for the future depending on the perceived severity of the event.
- Investigation of problem – An experienced utility employee is dispatched to investigate the problem. The employee identifies the nature of the problem (i.e., water main break causing property damage, service line leak, damaged hydrant, etc.), and initiates remedial action based on the severity of the incident.
- Remedial action – The solution to the problem may involve a temporary repair, a permanent repair or replacement, or notification of third party (for example, for service line repair outside the utility's responsibility).
- Follow-up – Follow-up steps may include recording crew times and other information related to the costs associated with the event, updating maintenance histories, alerting customers to the status, etc. At this time it is also important to collect and record other data related to the water main break, but not available from the field crews. Examples of these types of data include the age of the water main involved, the maintenance history for that water main, etc.

As noted earlier, at each step in the process different types of valuable data can be collected for later use. Figure 3.1 illustrates this concept and indicates the types of data that can be expected.

RECOMMENDED DATA COLLECTION

Most water utilities already have procedures and forms in place for recording basic information related to main breaks. Examples of actual forms used by a number of utilities are provided in Appendix B. In some cases, these forms are used exclusively by crews to compile field data. In other cases, the forms are more comprehensive, and both field and office data are compiled on the same form.

Table 3.1 and Figure 3.2 provide a comprehensive list of data and a form to capture the field and office water main break data identified as most useful for water utilities to use in developing water main renewal programs. A description of each data item requested is provided in Table 3.1. Note that some items may not be applicable to every water utility.

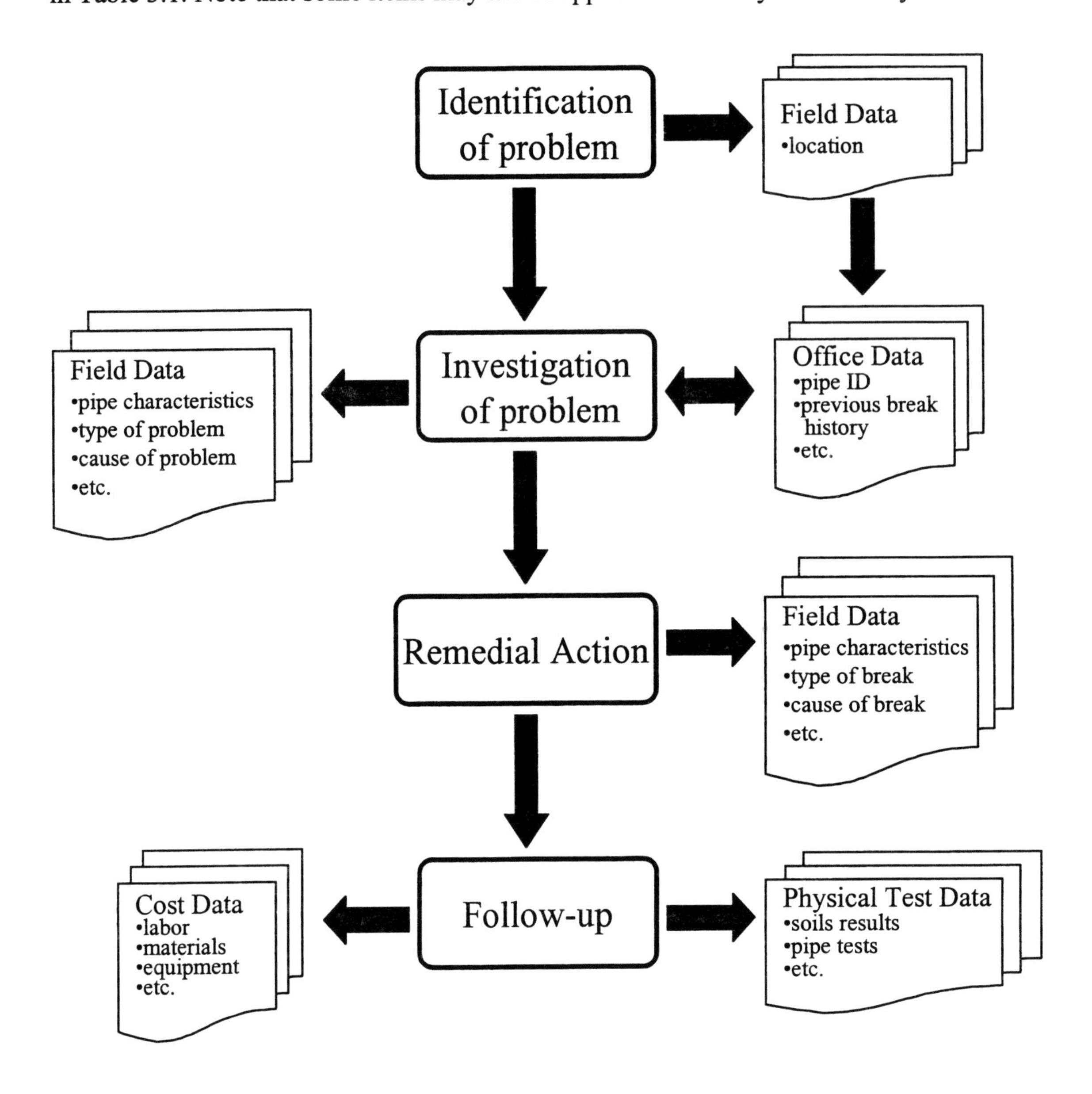

Figure 3.1 Data generated at each main break response step

Table 3.1

Recommended water main break data collection

Field	Description	Purpose
1) Date of excavation	The date on which the water main was exposed to repair the main break	Temporal analyses of main break trends
2) Employee name	The name of the utility employee who provided the field data	Follow up of questions
3) Address	The nearest address to the location of the break	Spatial analysis of main breaks
4) Nearest cross street	The cross street closest to the main break.	Accurate location of main breaks where street address is not applicable
5) Distance (ft) from cross street	The distance from the main break to the nearest cross street	Accurate location of main breaks where street address is not applicable
6) Pipe material	The material of the pipe	Trend analyses of main breaks by pipe material
7) Pipe diameter	The nominal diameter of the pipe	Trend analyses of main breaks by pipe diameter
8) Pipe protection	This is an indication of the corrosion protection characteristics of the pipe. Ductile iron pipe may be wrapped or unwrapped depending on when it was installed and local soil conditions. Other cathodic protection measures (anodes, etc.) should be noted.	Evaluation of corrosion protection programs
9) Joint type	The type of joint connecting pipe segments	Trend analyses of main breaks by joint type

(continued)

Table 3.1 (continued)

Field	Description	Purpose
10) Condition of pipe exterior	The extent and degree of external corrosion on the pipe, based on a visual inspection,	Provides a qualitative assessment of external corrosion. Can be used to identify problematic soil areas in the distribution system.
11) Condition of pipe interior (unlined pipe)	The extent and degree of internal corrosion of the unlined pipe, based on a visual inspection, check the boxes that best describe it. If the pipe interior was not observed (for example, if a pipe clamp was applied) select "Not observed".	Provides a qualitative assessment of internal corrosion. Can be used to assess the need for cleaning and lining.
12) Condition of pipe interior (cement lined pipe)	The condition of the cement lining, based on a visual inspection. If the pipe interior was not observed (for example, if a pipe clamp was applied) select "Not observed".	Provides a qualitative assessment of cement lining. Can be used to assess the need for relining or replacement.
13) Surface and traffic	Describes the surface under which the pipe is laid in order to estimate the traffic load the pipe experiences. "Roadway heavy (commercial) traffic" refers, for example, to a roadway that carries both a high volume of traffic, as well as a large number of trucks. "Roadway medium (mixed) traffic" would be indicative of a commercial shopping district where traffic volume might be high, but it consists primarily of automobiles. "Roadway light (residential) traffic" refers to a residential area where the traffic is almost exclusively automobile.	Helps to determine the cause of the main break. Can be used to examine trends in main breaks. May help in designing future water main installations. Data can be used in mechanistic models for predicting future main breaks.

(continued)

Table 3.1 (continued)

Field	Description	Purpose
14) Bedding	The "Type" and "Condition" of the bedding, based on a visual observation. It is recognized that the break itself is likely to disturb the bedding, and thus it is difficult to make observations about the bedding condition. Nevertheless, by examining the walls of the excavation in areas that are undisturbed, valuable information may be obtained.	Helps to determine the cause of the main break. May help in designing future water main installations.
15) Depth of pipe (surface to top of pipe)	The distance from the ground surface to the top of the pipe.	Helps to determine the cause of the main break. Can be used to examine trends in main breaks. May help in designing future water main installations. Data can be used in mechanistic models for predicting future main breaks.
16) Frost depth (from ground surface)	The depth of frost penetration from the ground surface at the time of the excavation	Helps to determine the cause of the main break. Can be used to examine trends in main breaks. May help in designing future water main installations. Data can be used in mechanistic models for predicting future main breaks.
17) Type of failure	The type of water main break, based on visual a observation	Allows for analyses of main break trends.
18) Show location of the failure	Illustration of where the failure occurred	Helps to determine the cause of the main break.
19) Probable cause of the failure	The most likely cause of the main break, based on a visual observation	Can be used to examine trends in main breaks. May help in designing future water main installations.

(continued)

Table 3.1 (continued)

Field	Description	Purpose
20) For pipe failures caused by bedding erosion, estimate the unsupported span length	An estimate of the length of pipe that was unsupported as a result of the erosion of the bedding under the pipe prior to the break. This should only be provided when the most likely cause of the break was beam failure. It is recognized that the bedding will likely have eroded further as a result of water flowing from the broken pipe, so field judgment will be required to estimate the initial unsupported length (i.e., prior to the break).	Helps to determine the cause of the main break. Used in mechanistic model to calculate loads on the pipe.
21) Soil category	The type of soil surrounding the pipe	Useful for analyses of main break trends. Helps to define characteristics of soil surrounding the pipe, which are needed for mechanistic modeling.
22) Soil temperature	The temperature of the soil at the depth of the pipe	Useful for analyses of main break trends. Helps to define soil characteristics surrounding the pipe, which are needed for mechanistic modeling.
23) Soil pH	The pH of the soil in the vicinity of the main break	Useful for analyses of main break trends. Helps to define soil characteristics surrounding the pipe, which are needed for mechanistic modeling.
24) Soil moisture content	The moisture content of the soil in the vicinity of the main break	Useful for analyses of main break trends. Helps to define soil characteristics surrounding the pipe, which are needed for mechanistic modeling.
25) Soil resistivity	The resistivity of the soil in the vicinity of the main break	Useful for analyses of main break trends. Helps to define soil characteristics surrounding the pipe, which are needed for mechanistic modeling.

(continued)

Table 3.1 (continued)

Field	Description	Purpose
26) Pipe wall thickness	Thickness of the pipe wall, typically from a laboratory measurement.	Helps to determine the cause of the main break. Used in mechanistic model to calculate strength of the pipe.
27) Pipe modulus of rupture	Modulus of rupture of a pipe sample taken from the site of the main break. Laboratory measurement.	Used in mechanistic model to calculate strength of the pipe.
28) Pipe fracture toughness	Fracture toughness of a pipe sample taken from the site of the main break. Laboratory measurement.	Used in mechanistic model to calculate strength of the pipe.
29) Employee name	The name of the employee researching the pipe history and system information	Follow up questions
30)Water main break ID	A unique identifier assigned by the utility to track the incident and related data	Provides a common identifier for data analyses.
31) Water main ID	A unique identifier to track the pipe on which the break occurred. If the utility has a water main inventory, this is simply the water main ID from that inventory. If no such database exists, the utility could consider other applicable identifiers such as Main Extension Number or Project Number.	Provides a common identifier for data analyses.
32) Installation year	The year in which the pipe that failed was installed	Useful for analyses of main break trends.

(continued)

Table 3.1 (continued)

Field	Description	Purpose
33) Length	The length of the pipe ID that failed. This length refers to the length recorded in a water main inventory.	Needed for developing pipe specific replacement program.
34) Dates of previous breaks on the same water main ID	Research existing main break databases and paper records to determine other water main breaks which occurred on the same "pipe". In this case "pipe" refers to the length of pipe identified by the Water Main ID requested previously.	Frequency of and duration between breaks are key modeling parameters.
35) Typical pressure in area of break	Describes the typical pressure in the pipe that failed. Note any unusual pressure fluctuations prior to the discovery of the main break	Useful for analyses of main break trends. Also used in mechanistic model for estimating load on the pipe.
36) Typical flow in area of break	Describes the typical flows in the pipe that failed. Note any unusual flow changes (fire flows, etc.) prior to the discovery of the main break.	
37) Water temperature	The water temperature at the time the main break was discovered. Treatment plant temperature is adequate for this purpose.	Useful for analyses of main break trends. Also used in mechanistic model for estimating load on the pipe.
38) Air temperature	The air temperature at the time the main break was discovered.	
39) # of consecutive days below 32°F	Indicate the number of consecutive days that the air temperature did not rise above 32°F prior to the discovery of the main break.	Used in mechanistic model for estimating load on the pipe.

(continued)

Table 3.1 (continued)

Field	Description	Purpose
40) Crew size	The number of employees responding to the main break	Used to estimate the labor cost.
41) Time on site	The duration of time needed to respond to the main break	Used to estimate the labor cost.
42) Was water service to customers interrupted?	Yes/no response to indicate if customers were without water as a result of the main break and subsequent repair activities	Used to estimate indirect cost of main break to customers.
43) Estimated number of customers whose service was interrupted	Estimated number of various types of customers that were without water as a result of the main break and subsequent repair activities	Used to estimate indirect cost of main break to customers.
44) Estimated time without service (hours)	Estimated length of time that various types of customers were without water	Used to estimate indirect cost of main break to customers.
45) Property damage	Yes/no response to indicate if any property damage occurred as a result of the main break.	Used to estimate the total costs of the main break.

WATER MAIN BREAK DATA COLLECTION FORM

1) Date of excavation: ______________________

2) Employee name: ______________________

3) Address: ______________________

4) Nearest cross street: ______________________

5) Distance (ft) from cross street: ____________

6) Pipe Material
- ❑ Unlined Cast Iron
- ❑ Cement Lined Cast Iron
- ❑ Ductile Iron
- ❑ Other (____________)

7) Pipe Diameter
- ❑ 6-inch
- ❑ 8-inch
- ❑ 10-inch
- ❑ 12-inch
- ❑ Other (__________)

8) Pipe Protection
- ❑ None seen
- ❑ Cathodic
- ❑ Wrapped
- ❑ Other (__________)

9) Joint Type
- ❑ Rigid (lead, leadite, etc.)
- ❑ Flexible
- ❑ Unknown

10) Condition of Pipe Exterior

Extent of Corrosion
- ❑ None
- ❑ Local (no pattern, distinct points)
- ❑ Uniform (over entire area)

Degree of Corrosion
- ❑ Negligible (0% - 20%)
- ❑ Light (20% - 50%)
- ❑ Moderate (50% - 80%)
- ❑ Severe (80% or more)

11) Condition of Pipe Interior (unlined pipe)

Extent of Corrosion
- ❑ None
- ❑ Local (no pattern, distinct points)
- ❑ Uniform (over entire area)
- ❑ Not observed

Degree of Corrosion
- ❑ Negligible (0% - 20%)
- ❑ Light (20% - 50%)
- ❑ Moderate (50% - 80%)
- ❑ Severe (80% or more)
- ❑ Not observed

12) Condition of Pipe Interior (cement lined pipe)
- ❑ Lining intact and in good condition
- ❑ Lining cracked or missing
- ❑ Not observed

13) Surface and Traffic
- ❑ Unpaved
- ❑ Sidewalk
- ❑ Easement – no traffic
- ❑ Roadway – light (residential) traffic
- ❑ Roadway – medium (mixed) traffic
- ❑ Roadway – heavy (commercial) traffic

14) Bedding (make observations at face of excavation in undisturbed areas)

Type
- ❑ Granular
- ❑ Sand
- ❑ Native soil

Condition
- ❑ Obvious voids or washout
- ❑ Appears uniform

15) Depth of Pipe (surface to top of pipe) ________ **feet**

16) Frost Depth (from ground surface) ________ **inches**

17) Type of failure
- ❑ blow out
- ❑ circumferential break
- ❑ longitudinal split along pipe
- ❑ split bell/joint
- ❑ corrosion hole
- ❑ other (____________)

18) Show location of the failure (not necessary for circumferential break)

19) Probable cause of the failure:
- ❑ settlement
- ❑ pipe/rock contact
- ❑ bedding erosion
- ❑ frost load
- ❑ traffic load
- ❑ high pressure
- ❑ water temperature change
- ❑ third party
- ❑ corrosion
- ❑ unknown

Did corrosion protection fail?
- ❑ Yes
- ❑ No
- ❑ Unknown

20) For pipe failures caused by bedding erosion, estimate the unsupported span length:

____________ **feet**

Figure 3.2 Data collection form for water main breaks

Physical Testing Data

21) Soil Category

- ❑ Sand
- ❑ Silt
- ❑ Clay
- ❑ Rock
- ❑ Other (________________)

22) Soil Temperature ______________

23) Soil pH ______________

24) Soil Moisture Content ______________

25) Soil Resistivity ______________

26) Pipe wall thickness ______________

27) Pipe modulus of rupture ______________

28) Pipe fracture toughness ______________

Water Main Break History and System Information

29) Employee Name ________________________
(completing this portion of the form)

30) Water Main Break ID __________________

31) Water Main ID __________________

32) Installation Year ____________

33) Length ____________

34) Dates of prior breaks on the same water main ID

__________________ __________________
__________________ __________________
__________________ __________________

35) Typical Pressure in Area of Break

- ❑ 20-40 psi
- ❑ 40-80 psi
- ❑ 80-100 psi
- ❑ 100 psi

Note any unusual changes in pressure in the 24 hours prior to the break. Include any valve/hydrant/pump operations in the area in the 24 hours prior to the break.

36) Typical Flow in Area of Break

- ❑ 0-100 gpm
- ❑ 100-500 gpm
- ❑ 500-1,000 gpm
- ❑ 1,000 gpm

Note any unusual changes in flow in the 24 hours prior to the break.

37) Water Temperature ________ °F or ________ °C

38) Air Temperature ________ °F or ________ °C

39) # of consecutive days below 32 °F (0 °C) __________

Water Main Break Cost Data

40) Crew size __________

41) Time on site _____ **hours**

42) Was water service to customers interrupted?

- ❑ Yes
- ❑ No

43) Estimated number of customers whose service was interrupted

residential _______
commercial _______
industrial _______
schools _______
hospitals _______
other _______

44) Estimated time without service (hours)

residential _______
commercial _______
industrial _______
schools _______
hospitals _______
other _______

45) Property damage

- ❑ Yes
- ❑ No

Figure 3.2 (continued)

Field Data

Field data refers to information that can only be collected at the site of the water main break. This includes all visual observations of the failed water main, its environment, and the steps taken to repair it. Some of the requested information is obvious, and only requires that the field crews record it on the form. Other requested information requires some subjective judgment on the part of the field crew to assess the condition of the water main and the potential cause of the main break.

Field crews responsible for responding to water main breaks should receive training on completing the data collection form. The training should include a description of data requests that require subjective judgment on the part of the field crew. For example, item 10 on Figure 3.2 requests an assessment of the exterior condition of the pipe. Each utility should develop its own criteria for assessing the pipe, and examples of pipe sections fitting each criterion should be shown in the training as examples. Providing examples like this helps to ensure that the data that is collected is consistent even though multiple crews may be collecting the data. Finally, the training should stress the importance of the data collection effort. The data collection should be presented as a critical component of the utility's overall infrastructure renewal program.

Ideally, all data collected would be based on measurements of physical properties or other objective criteria. This would ensure consistency of data and allow the utility to easily compare and evaluate main break data over time. It would also allow the data to be shared with other utilities. This would be done, for example, for the purpose of examining general causes of main breaks, and identifying potential remedies to the problem of main breaks. However, utilities should not abandon collection of certain field data simply because subjective judgment must be applied. For example, in the suggested data collection practices (Table 3.1) items 10, 11, and 12 request subjective assessments of the pipe's interior and exterior condition. Obviously, field crews' perception of condition will vary within a utility, and these data are not likely to be useful to other utilities. However, if these subjective ratings of the pipes' condition were to be plotted on a map, they could very well indicate areas where the pipes consistently received a "poor" condition rating. This could lead to a water main renewal program targeted at these areas, or at least focus further investigations towards those areas. This illustrates the importance of collecting information, even if it is not objectively based.

Office Data

Office data includes information related to the water main that has failed that cannot be observed in the field. It also includes data that might be available to field crews, but is more easily and efficiently accessed by office staff. This includes operational parameters (pressure and flow), environmental conditions (air and water temperatures), and a history of the pipe that failed. These requested data may or may not be available to the field crews that complete the field data portion of Figure 3.2.

An important example of office data is the identification of the specific pipe that failed. Field crews can provide the location of the main break in terms of street address or by referencing cross streets. However, in order to thoroughly analyze main break trends it is important to have the installation and maintenance history of the pipe available. Once the field data for a main break event has been submitted, the associated pipe ID should be identified and recorded. Ideally, it would then be possible to investigate the history of the pipe.

This highlights the need for a system to specifically identify a pipe section. Some utilities use the installation record of a pipe as its identifier. For example, all pipe installed under a single contract number would share the same pipe ID. This allows the utility to reference pipe installation information, but if the length of pipe represented by the pipe ID is too long, it may be difficult to isolate specific problems. Other utilities assign pipe IDs using combinations of map or drawing sheet number, street, and even block along a street. This provides a unique identifier for a reasonable length of pipe, but considerable work may be required to then link these pipe IDs back to pipe installation and maintenance histories. A GIS provides the most powerful method of managing, retrieving, and analyzing pipe specific information.

Physical Testing Data

In addition to the field and office information that should be collected every time a water main breaks, there are other more detailed physical data that can prove valuable. Testing of pipe and surrounding soils can provide critical information for a utility in developing a water main renewal program. For example, a utility experiencing reduced flows and water quality problems related to the tuberculation of cast iron water mains may consider cleaning and cement mortar lining as a solution. However, if the existing cast iron pipe is not structurally sound, replacing the pipe may be a more appropriate solution. Testing of soils is critical in the development of corrosion protection programs to prolong the life of pipe. A water main renewal program that does not adequately address soil corrosivity may not achieve the desired results. Lack of data regarding soil conditions could lead to premature aging of water mains installed as part of the renewal program.

Pipe Testing

The physical testing described here is appropriate for cast iron pipes. All structural testing of pipe should be performed following the specifications of the American Society for Testing and Materials (ASTM) for overall methodology, and following AWWA standards for water pipe specific requirements. The governing standards for cast iron pipe are AWWA C106/ANSI A21.6 for metal mold manufacturing and AWWA C108/ANSI A21.8 for sand mold manufacturing. For ductile iron pipe the governing standard is AWWA C151/ANSI A21.51.

The following tests could provide valuable information about the condition of a pipe:

- Pipe wall thickness (remaining) – Remaining wall thickness is a measure of loss of metal due to corrosion, and provides an indication of remaining structural strength. Determination of wall thickness is by an ultrasonic thickness gauge according to ASTM E-797. Alternatively, a micrometer can be used to take representative measurements around the circumference of the pipe.
- Tensile strength – The general testing methodology is specified in ASTM E-8 for both cast and ductile iron. Preparation of sample coupons differs for cast and ductile iron, and the appropriate AWWA/ANSI specifications should be followed.
- Ring modulus – This measures the pipes resistance to crushing forces.
- Modulus of Rupture - The modulus of rupture test should be done in accordance with AWWA C106/108, which also specifies the number of samples and sample preparation.
- Hardness – Rockwell hardness is specified only for metal mold cast iron, and sample preparation and test methodology is provided in ASTM E-18.

- Carbon Percentage – The percent carbon in the sample is a measure of the extent of graphitization of the pipe. Graphitization is essentially the leaching of iron from the pipe, resulting in reduced pipe strength. Determination of the carbon percentage in the pipe can be done using the Leco Carbon and Sulfur Analyzer.
- Fracture toughness – This test is a measure of the pipe strength resulting from the weakening produced by corrosion pits. Details of testing the fracture toughness of cast iron pipes can be found in Rajani et al. (2000).

The costs of conducting all of these tests on every pipe sample could be prohibitive. Some of these tests have been determined to be critical in the use of mechanistic models of main failures, while others are useful but not critical. In order to minimize costs, only the following critical tests are recommended:

- Wall thickness
- Modulus of rupture
- Tensile strength

Wall thickness is considered to be the most important parameter to measure, and is probably the easiest for a utility to perform in-house. It is not physically or economically practical to take a pipe sample for material testing from every water main break. For example, many utilities rely on pipe clamps to repair circumferential breaks, and therefore, do not remove a piece of the pipe. Ideally, the samples that are collected would be equally representative of the type of pipe in the system. That is, samples should be collected from different groupings of pipe materials, diameters, and ages.

A minimum pipe length of four feet should be collected. For a circumferential break, this should include two foot on either side of the break. This will provide sufficient pipe material to test the physical properties of the pipe, including a portion that did not fail. For a longitudinal break, the pipe sample should be selected from a "typical" segment of the failed pipe, but should at least include some portion that did not fail.

Few water utilities have the in-house facilities to perform these tests, and consequently outside laboratories must usually be used. Some savings in cost could be achieved if the utility prepares the required specimens. Details of specimen preparation are provided in the appropriate ASTM, AWWA, or ANSI specifications. However, it should be noted that the specimen preparation is very exacting, and requires a skilled machinist with the appropriate equipment. Preparation of cast iron specimens is particularly difficult due to the brittle nature of cast iron. Therefore, it is recommended that the laboratory prepare the test specimens from pipe samples sent by utilities.

Samples sent to the laboratory should be labeled carefully so that the results can be matched to the water main break documentation previously described. The laboratory should be consulted for the appropriate chain of custody (COC) procedures. Each pipe should be identified using two methods so that it can be positively identified later. First, spray paint or other indelible marker should be used to directly label the pipe. Second, a tag should be attached to the pipe sample with wire. The painted label and tag should include the following information as a minimum:

- Utility name
- Main break ID (from the data collection form)
- Date of main break

If the laboratory COC procedures require additional information, it should be provided. Finally, the pipe should be marked to indicate its orientation in the ground. The top of the pipe should be indicated with paint, and the bell and spigot ends should also be marked.

Soil Testing

Soil characteristics can impact the condition of cast iron and ductile iron water mains. In particular, corrosive soils can greatly reduce the life span of these pipes unless precautions are taken. Encasement of ductile iron pipe in a polyethylene wrap has become standard practice for many utilities in order to protect against corrosive soil. ANSI/AWWA C105/A21.5-82 defines the standards for polyethylene encasement of ductile iron pipe. This same standard also describes soil-testing procedures to determine if polyethylene encasement is needed to protect against corrosive soil. Thus, ANSI/AWWA C105/A21.5-82 is a good guide to assist utilities in developing soil-testing programs.

The following eight parameters are suggested for evaluating polyethylene encasement requirements according to Appendix A of ANSI/AWWA C105/A21.5-82:

1. Earth (soil) resistivity – Three methods for determining earth (soil) resistivity include four-pin, single-probe, and soil box. Resistivity is highly dependent on soil moisture content and temperature. Therefore, it is recommended that interpretation of results be based on the lowest reading obtained.
2. pH – A direct reading of soil pH is made, accounting for soil pH.
3. Moisture content – Soil moisture content is accounted for to some extent under earth resistivity, but is so important to overall corrosion that it also needs to be addressed separately. Conditions typically vary widely over the course of the year. Therefore, a general description of the moisture content can be provided as 1) poor drainage, continuously wet, 2) fair drainage, generally moist, or 3) good drainage, generally dry.
4. Soil description (category)– Several basic soil types should be noted, including sand, loam, silt, or clay. Unusual soils such as peat should also be noted. More detailed soil testing, such as particle size, plasticity, friability, and uniformity, can also be performed.
5. Oxidation-reduction (redox) potential – Redox potential is significant because the most common sulfate-reducing bacteria live only under anaerobic conditions. Soil samples can undergo a change in redox potential on exposure to air, and thus should be tested immediately.
6. Sulfides – A positive sulfide reaction indicates a potential problem due to sulfate reducing bacteria.
7. Potential stray direct current – The proximity of potential sources of stray current to the pipe in question should be noted. Potential sources include other utility's cathodic protection systems, electric railways, and industrial equipment (including welding).
8. Experience with existing installations in the area – Possibly the most important indication of soil corrosivity is past experience in the area. This again highlights the need to collect and maintain information related to main breaks that would indicate external corrosion.

Discussions with other utilities, particularly gas utilities, can also provide useful information regarding potential corrosion problems.

The first six parameters require testing of the soil. These tests are preferably performed in the field, as changes in soil characteristics can occur during transport to a laboratory. The remaining two parameters do not require testing, but should be noted each time a water main break is investigated. Eventually, a database of soil characteristics based on actual experience can be developed, providing important information for evaluating the existing condition of water mains and planning for future distribution system renewal.

An additional soil characteristic determined to be critical to the modeling of main breaks is soil temperature. Soil temperature is needed to calculate the thermal stresses experienced by the pipe prior to failure. Correlations between ambient air temperature and soil temperature have been investigated, but found to be dependent on too many other factors to be practically applied. Therefore, it is recommended that soil temperatures be collected at the time of the break excavation.

In summary, the following soil characteristics were identified as critical in this project and were included in Figure 3.2:

21) Soil category
22) Soil temperature
23) Soil pH
24) Soil moisture content
25) Soil resistivity

All of these soil characteristics can be measured in the field. Soil categories can be broad and field personnel can be trained to recognize and record them. Soil temperature can be measured using a thermometer placed into the wall of the excavation at the depth of the pipe. It is recognized that the reading may be subject to ambient air temperatures, but it is believed that a valid reading can be made. Inexpensive instruments are commercially available for measuring soil pH and soil moisture content.

Soil resistivity measurements are more complicated than the others, but there are water utilities that have collected resistivity readings. Both the City of Calgary and the RMOC have collected soil resistivity data in the field using utility personnel.

Testing of soils in conjunction with the response to a water main break is logical since the soils around the pipe are exposed. Repairing a break may be the only time that the soil has been accessible since the pipe was originally installed. However, there are valid reasons why soil sampling during main break repair might be less than desirable. First, one of the primary soil characteristics to sample is soil moisture. Obviously, water from the broken main will impact the soil moisture value. It may or may not be possible to obtain a representative sample of ambient soil moisture from some area of the excavation away from the break. Secondly, the use of soil testing equipment will require the cooperation of field crews trained to use the equipment or the assistance of other utility personnel trained to conduct soil tests.

For these reasons a soil characterization program conducted separately from main break repairs may be more practical and successful for utilities. A general understanding of soil conditions throughout a distribution system can be obtained from soils maps prepared by government agencies as well as from experienced utility personnel. Targeted sampling of known or suspected problem areas (i.e., corrosive soils) could then be undertaken under controlled conditions (i.e., not during main break repair).

If soil measurements cannot be made in the field, then the following soil testing procedure developed by the National Research Council (NRC) Canada can be used (Rajani et al. 2000).

1. Before taking soil samples from the trenches, a utility soil engineer should become familiar with Appendix A of ANSI/AWWA C105/A21.5-88.
2. Soil samples should be taken at four selected sites of longitudinal split and circular main breaks. All relevant soil data indicated in Figure 3.2 should be completed.
3. Two soil samples from every trench, on from the bell end and one from the spigot end of the trench, should be collected. The samples should be taken from the sides of the trench as close to the pipe as possible in undisturbed soil. Soil samples can

be damp or moist but must not be waterlogged or saturated. If the soil closest to the pipe is too wet, the closest good sample should be taken.

4. Each soil sample from the trench should be placed in its own polyethylene bag. A clean shovel should be used to collect the samples. Each sample should weigh at least 3 to 5 kg (6 to 10 lb). The sample bags should be securely sealed, and as much air as possible should be removed. Air may cause unreliable soil test results.
5. Each soil sample should be clearly and securely labeled. Laboratory COC guidelines should be followed, but as a minimum the label should indicate the following:

   - Utility name
   - Date of sample
   - Time of sample
   - Address
   - Water main break ID

6. One soil sample bag from each trench should be sent to the soil laboratory for soil classification, index and soil resistivity measurements (soil-box).

Cost Data

Cost data should capture the true costs of responding to and repairing a water main break. The direct costs, including labor and materials, are straightforward and usually recorded by some group within the water utility. Other information related to the indirect costs of water main breaks is not typically recorded by utilities. Utilities should begin to collect these data, and ultimately use them in developing water main renewal programs.

FIELD TEST

The data collection recommendations described in Table 3.1 were used by four participating utilities to record water main break events over an approximately 6-month long period. The implementation of these recommendations highlighted several important points for a utility to consider when designing its own data collection program. These include:

1. Field crew input is vital to the success of the effort. The primary function of the field crew when there is a water main break is to get the pipe back into service as soon as possible. This may involve a simple repair or a partial pipe replacement. In any event, the field crew may be reluctant to collect and record data that is viewed to be extraneous to this basic function. The utility staff that is requesting the data (engineering, management, etc.) must work with the field crews to explain the importance of the data collection effort. They should stress that the goal of such a program is to ultimately reduce the number of main breaks that the crew must deal with.
2. Data collection forms must be customized for each utility. The forms shown in Figure 3.2 and in Appendix B provide examples of how data collection forms could be organized. However, each utility has unique terminology that utility personnel are accustomed to using. Every effort should be made to use utility-specific terminology.
3. The data collection form should be kept as simple as possible. Check boxes and yes/no type responses should be used whenever possible. This makes it easier and quicker for the field crews to complete the form. This should increase the number of forms that are completed and returned, thus improving the thoroughness of the data collection. Standardizing responses through the use of check boxes also facilitates data analysis.
4. Handling of soil and pipe samples must be considered before the samples are collected. Soil and pipe samples are likely to be collected only from selected main break locations. Procedures must be in place to properly label, record, and deliver the samples to the appropriate laboratory before the sampling takes place. Otherwise, samples could be lost or rendered unusable. As an example, in this project problems arose at the pipe-testing laboratory because several pipe samples were not adequately labeled. This made it difficult in several instances to coordinate laboratory results with main break locations.

5. The coordination of data collection must be assigned to a specific individual. In this project, two of the utilities used university students to compile the main break data. Assigning the responsibility of water main break data collection and management to a single person (or group) ensures that the task will receive the attention needed.

In this project, the collection of water main break data began in January 1999. Ultimately, PSWC, PWB, RMOC, and UWNJ provided data. Table 3.2 summarizes the data collection effort. Laboratory results for the pipe and soils testing are provided in Table 3.3.

Table 3.2

Summary of water main break data collection effort

Utility	Number of water main breaks	Number of pipe and soil samples collected
PSWC	101	14
PWB	8	4
RMOC	118	19
UWNJ	227	7

DATA MANAGEMENT

Collecting the appropriate water main break data is only the first step in the use of these data to develop water main renewal programs. Some utilities have been collecting water main break data for many years. In some cases these utilities have only the most basic information recorded. However, due to the long period of time covered by the data, valuable input to the water main renewal decision-making process can be provided. If the data are only available in paper records and not organized in a logical fashion, then the usefulness of the data might be limited.

Utilities should plan for management of the water main break data as carefully as they plan for its collection. In the simplest form, a computer database of water main breaks should be created. This database should include fields for all of the data that a utility is or will collect. Ideally, the database would be linked to a computerized water main inventory. One of the recommended (see Figure 3.2) "Office Data" fields is Pipe ID. Using the same Pipe ID in both the water main break database and the water main inventory facilitates data analysis.

More sophisticated data management systems would link customer service information systems, water main inventories, break databases, and work management systems (WMS) to provide a comprehensive data set for analysis. The customer information system would capture the initial identification of the problem, the water main inventory and break databases would provide the specific information related to the pipe and the break events, and the WMS would provide detailed cost information.

Table 3.3

Summary of laboratory testing results

Sample	Material	Diameter	Installation Year	Pressure (psi)	Wall thickness (in)	Modulus of rupture (psi)	Fracture toughness (Mpa*m1/2)	pH	Resistivity (ohm-cm)	Moisture content
RMOC 1	Lined cast iron	6"	1961	75	0.466	32,045	11.4	5.7	1,193	32%
RMOC 2	Lined cast iron	8"	1962	62	0.451	47,705	11.8	7.1	2,987	39%
RMOC 3	Unlined cast iron	4"	1950	71	0.295	41,905	-	7.2	4,040	17%
RMOC 4	Unlined cast iron	16"	1954	42	0.747	48,302	12.0	7.2	179	68%
RMOC 5	Unlined cast iron	6"	1954	65	0.446	34,475	11.1	7.0	2,025	16%
RMOC 6	Unlined cast iron	8"	1960	62	0.559	51,932	12.2	7.3	882	35%
RMOC 7	Lined ductile iron	8"	1974	85	0.344	76,828	-	-	-	-
RMOC 8	Unlined cast iron	8"	1960	63	0.491	13,943	12.2	6.9	819	46%
RMOC 9	Unlined cast iron	6"	1977	58	0.407	48,430	13.0	5.8	216	29%
RMOC 10	Lined cast iron	8"	1971	57	0.433	53,070	13.8	5.4	1,119	38%
RMOC 11	Lined ductile iron	6"	1974	64	0.293	74,095	-	-	-	-
RMOC 12	Lined ductile iron	6"	1972	80	0.287	77,575	-	6.0	2,749	35%
RMOC 13	Lined ductile iron	12"	1977	68	0.267	78,880	-	7.3	321	26%
RMOC 14	Lined ductile iron	8"	1973	83	0.221	123,540	-	7.1	709	36%
RMOC 15	Unlined cast iron	6"	1958	70	0.325	50,895	14.2	7.4	2,172	32%
RMOC 16	Unlined cast iron	6"	1961	62	0.436	47,850	11.9	7.4	990	35%
RMOC 17	Unlined cast iron	6"	1961	62	0.508	41,905	11.2	6.3	861	35%
RMOC 18	Unlined cast iron	6"	1955	59	0.440	55,245	13.0	-	-	-
RMOC 19	Lined cast iron	6"	1963	48	0.330	-	11.9	7.1	1,583	59%
Portland 1	Lined cast iron	6"	1960	40-80	0.350	20,485	90.3	6.1	-	-
Portland 2	Lined cast iron	6"	1960	40-80	0.388	17,332	-	6.1	-	-
Portland 3	Lined cast iron	4"	-	80-100	0.405	9,662	83.5	-	-	-
Portland 4	Lined cast iron	4"	-	40-80	0.514	10,835	87.7	-	-	-
Portland 5	Unlined cast iron	8"	1953	80-100	0.449	21,324	99.9	5.8	-	-
PSWC 1	Unlined cast iron	6"	1948	> 100	0.437	23,589	100.5	7.7	10,780	18.4%
PSWC 2	Lined cast iron	6"	1950	> 100	0.416	9,815	-	8.0	3,590	30.6%
PSWC 3	Lined cast iron	6"	1923	-	0.464	11,415	77.1	8.0	4,180	32.1%
PSWC 4	Unlined cast iron	8"	1951	> 100	0.424	18,618	-	7.2	3,080	31.1%

(continued)

Table 3.3 (continued)

Sample	Material	Diameter	Installation Year	Pressure (psi)	Wall thickness (in)	Modulus of rupture (psi)	Fracture toughness (Mpa*m1/2)	pH	Resistivity (ohm-cm)	Moisture content
UWNJ 1	Unlined cast iron	8"	1921	80-100	0.467	-	-	6.6	1,860	35.9%
UWNJ 2	Unlined cast iron	6"	-	80-100	0.396	-	-	8.2	2,770	27.5%
UWNJ 3	Unlined cast iron	6"	-	80-100	0.461	8,221	-	8.2	4,230	27.3%
UWNJ 4	Unlined cast iron	6"	-	80-100	0.365	-	-	7.9	1,170	22.1%
UWNJ 5	Unlined cast iron	6"	-	80-100	0.408	6,768	17.8	7.1	1,870	17.2%
UWNJ 6	Unlined cast iron	8"	1921	80-100	0.466	8,503	61.4	6.6	1,860	35.9%
UWNJ 7	cast iron	6"	-	80-100	0.395	10,119	-	7.9	1,170	22.1%
UWNJ 8	cast iron	6"	-	80-100	0.303	8,538	-	7.6	2,200	27.7%
UWNJ 9	cast iron	6"	-	80-100	0.442	9,123	-	8.2	4,230	27.3%
UWNJ 10	cast iron	6"	1961	80-100	0.393	19,776	-	6.8	1,370	26.8%

CHAPTER 4
APPLICATIONS OF WATER MAIN BREAK DATA

One of the goals of this project was to develop tools for planning water distribution system renewal using data collected in conjunction with water main breaks as a key component. Chapter 3 described the types of data to be gathered and procedures for collecting and managing that data. This chapter describes the use of water main break data for developing prioritized water main renewal programs. A mechanistic model for prioritizing water main renewals on the basis of vulnerability to failure was developed for this project and is described in this chapter. Simpler uses of water main break data are first presented as a guide for utilities that cannot take advantage of this mechanistic model due to input data limitations.

The availability of historic water main break data provides a utility with much of the critical data needed to develop an effective water main renewal program. Various types of data analyses yield different information about the condition of pipes in the system, and about the most effective means of renewing the pipes. Even simple trend analysis can provide insight into the causes of main breaks. The results of these analyses could also suggest renewal strategies that may not be readily apparent.

A variety of applications using water main break data are presented in the following sections. The data needed to perform these applications are listed and described, and examples are provided using hypothetical and actual water main break data. A portion of a hypothetical main break database is provided in Table 4.1. Fields in the database are defined as follows:

- Pipe ID – a unique identifier for a specific pipe
- Installation Year – the year that the pipe was installed
- Pipe Diameter – the diameter of the pipe (in.)
- Pipe Material – the material of the pipe (PCI for pit cast iron and SCI for spun cast iron)
- Joint Type – the type of joint connecting two pipe sections (R for rigid and F for flexible)
- Date of Break – the date on which the break occurred
- Type of Break – the type of failure (CB for circumferential break and LB for longitudinal break)
- Grid Location – defines the location of the break according to a map grid

Table 4.1

Portion of a hypothetical main break database

Pipe ID	Installation Year	Pipe Diameter	Pipe Material	Joint Type	Date of Break	Type of Break	Grid Location
1088	1912	20	PCI		1/15/89	LB	22L21
1088	1912	20	PCI		4/2/90	LB	23L53
1088	1912	20	PCI		11/5/91	LB	22L21
1088	1912	20	PCI		8/27/92	LB	24L51
1088	1912	20	PCI		3/29/95	CB	22L12
1088	1938	6	SCI		7/25/96	LB	23L51
1091	1938	6	SCI		8/20/97	LB	23M32
1109	1938	6	SCI		8/9/91	LB	19K61
1109	1938	6	SCI		12/25/91	CH	19K61
1109	1938	6	SCI		12/7/96	CH	19K61
1109	1938	6	SCI		10/11/97	CB	19K61
1112	1936	6	SCI		9/9/83	CH	15G42
1112	1936	6	SCI		9/16/83	LB	15G42
1112	1936	6	SCI		1/3/87	LB	15G44
1112	1936	6	SCI		1/3/87	CB	15G44
1112	1936	6	SCI		12/13/87	CH	15G42
1112	1936	6	SCI		1/4/89	CB	15G42
1112	1936	6	SCI		8/28/91	LB	15G42
1112	1936	6	SCI		9/10/96	LB	15G41
1113	1931	6	SCI		6/7/84	CB	15G53
1113	1931	6	SCI		6/8/84	LB	15G53
1113	1931	6	SCI		6/9/84	LB	15G53
1113	1931	6	SCI		6/9/84	LB	15G53
1113	1931	6	SCI		9/30/85	LB	15G44
1113	1931	6	SCI		12/14/87	LB	15G53
1113	1931	6	SCI		6/20/90	LB	15G53
1113	1931	6	SCI		10/25/90	LB	15G44
1113	1931	6	SCI		6/14/92	LB	15G53
1113	1931	6	SCI		7/10/92	LB	15G53
1113	1931	6	SCI		9/1/92	LB	15G53
1113	1931	6	SCI		3/9/93	CB	15G53
1113	1931	6	SCI		6/16/93	CH	15G53
1113	1931	6	SCI		12/13/95	LB	15G53
1113	1931	6	SCI		12/17/95	LB	15G53
1113	1931	6	SCI		7/30/97	CH	15G53
1113	1931	6	SCI		7/30/97	CH	15G53
1120	1923	6	PCI		9/16/92	LB	21K31
1121	1932	6	SCI		8/14/83	LB	18K63
1121	1932	6	SCI		1/22/85	CB	18K63
1121	1932	6	SCI		3/28/87	LB	18K63
1121	1932	6	SCI		10/22/89	CB	18K63
1121	1932	6	SCI		6/29/94	LB	18K63
1121	1932	6	SCI		6/29/94	CH	18K63
1122	1945	6	SCI		12/20/89	LB	23J44
1122	1936	6	SCI		1/1/91	CB	23J53

MAIN BREAK OCCURRENCE

One of the most straightforward methodologies for prioritizing water mains for renewal is to look at the break history for each pipe. Many utilities have established criteria for the number of breaks that they are willing to accept before a pipe is replaced. Other utilities use the number of main breaks occurring on a pipe as one parameter used to score a pipe in a priority ranking system. Break rates (breaks per mile per year, for example) can be calculated for different pipe segments and used as an index to prioritize water main replacements.

Using break data in this fashion provides a simple method for targeting specific pipes for renewal. However, this methodology is reactive (i.e., after the breaks occur) rather than proactive. The utility is allowing the pipe to fail, sometimes on several occasions, before any action is taken to renew it. A proactive approach requires the utility to at least consider the probable future of the pipe and to take pre-emptive action to avoid future main breaks.

The only data required for this type of simple analysis is a record of the pipe ID and when main breaks have occurred. These are included in Table 4.1 as Pipe ID and Date of Break, and were included in the recommended water main break data collection requirements provided in Table 3.1. The data in Table 4.1 shows the break history for eight Pipe IDs. Closer examination shows that two of the pipes (1091 and 1120) have failed only one time each over the period of time covered by the database. On the other hand, Pipe ID 1113 has failed a total of 17 times, and may have failed additional times prior to the time period covered by the database.

Simply knowing how many times a particular pipe has failed does not necessarily determine if or when that pipe should be renewed. In the Table 4.1 example, Pipe ID 1113 has failed 17 times over the course of 16 years, while Pipe ID 1112 has only failed 8 times over 17 years. However, one vital piece of data is missing from the analysis, namely the length of the pipe. If Pipe ID 1113 represented a pipe that was 2 miles long, the 17 breaks over 16 years represents a break rate equivalent to approximately 53 breaks/100 mi/yr. At the same time if Pipe ID 1112 was only 1,000 feet long, its 8 breaks equate to a rate of 470 breaks/100 mi/yr. While neither break rate is acceptable, it is clear that other factors besides simply the number of breaks must be considered.

MATERIAL ANALYSIS

An important factor to consider in assessing water main breaks and planning for water main renewal is pipe material. Most utilities are aware of general water main break patterns with respect to different pipe materials. For example, it may be obvious to the utility that cast iron pipes are failing more frequently than other pipe materials. However, more detailed analyses can sometimes pinpoint specific categories of cast iron pipe that are failing more frequently, and thus may be primary candidates for renewal.

At first glance it might seem obvious to target the oldest pipes in the system for renewal. One reason why it is not appropriate to simply follow this strategy is related to the history of pipe materials used in the water industry. The earliest cast iron pipe still in use was manufactured using a casting technique where molten iron was poured into molds. The resulting pipe, called pit cast iron, had relatively thick, but non-uniform, wall thickness. Spun cast iron pipe was introduced in the 1920's. This pipe was manufactured using centrifugal casting, resulting in a thinner and more uniform wall thickness. The thinner walls of the spun cast iron pipe has resulted in generally higher break rates for this type of pipe in most water utilities compared to pit cast iron pipe.

Two other material-related issues could impact water main breaks. First, pipe manufacturers and suppliers for the utility may have changed over time. Differences in the quality of materials may be reflected in main break rates. Second, the type of joint can affect pipe performance. Rigid joints, such as poured lead and leadite, restrict the movement of pipe and may lead to main breaks.

Each of these issues can be evaluated using main break data. By comparing the rate of main breaks for different pipe materials, including joint type, the utility can at least identify those general categories of pipe that should have a high renewal priority. By examining main breaks for different vintages (i.e., installation dates) of pipe, it may even be possible to identify periods when pipe quality was poor. Once again, these pipes could receive higher renewal priority.

Thus, the data required to conduct these types of material analysis include the standard pipe ID, and a record of when the pipe failed. Other data required would be the pipe material, joint type and installation year of the pipe. These latter data items are good examples of the office data described in Chapter 3, and illustrate the need to have a link between a main break database and a water main inventory.

SLCWC maintains a detailed water main break database. The principal pipe materials in their distribution system are pit cast iron, spun cast iron with rigid joints, spun cast iron with flexible joints, and ductile iron. Using its main break database SLCWC conducted an analysis of breaks by material. The results of the analysis are shown in Figure 4.1, which indicates that although the spun cast with rigid joint pipe represents approximately 25% of the total length of pipe in the system, this category of pipe has accounted for close to 60% of the main breaks since 1983. Conversely, spun cast iron pipe with flexible joints accounts for 42% of the system yet has only produced 24% of the main breaks. Thus, the primary category of pipes that should be targeted for renewal in the SLCWC system is clear.

SPATIAL ANALYSIS

One type of trend analysis to consider is the spatial distribution of water main breaks. Some utilities simply plot the locations of main breaks on a map. A similar, but more sophisticated, approach utilizes a GIS to track main break occurrences. For this type of analysis, the only type of water main break data required is the location of the break.

Once the locations of main breaks have been plotted, a review may indicate potential causes of the breaks. For example, if main breaks are concentrated near a pump station or within a particular pressure zone, high pressure or water hammer may be causing the breaks. In this case, modifications to the pump station or operating procedures could reduce the number of main breaks. As another example, soil conditions can have a significant impact on corrosion of the pipe wall and affect the structural condition of the pipe. High incidence of main breaks in a given area could prompt a utility to conduct a soil sampling program in that area. The results of the sampling would provide valuable information for the utility in planning for future water main renewal.

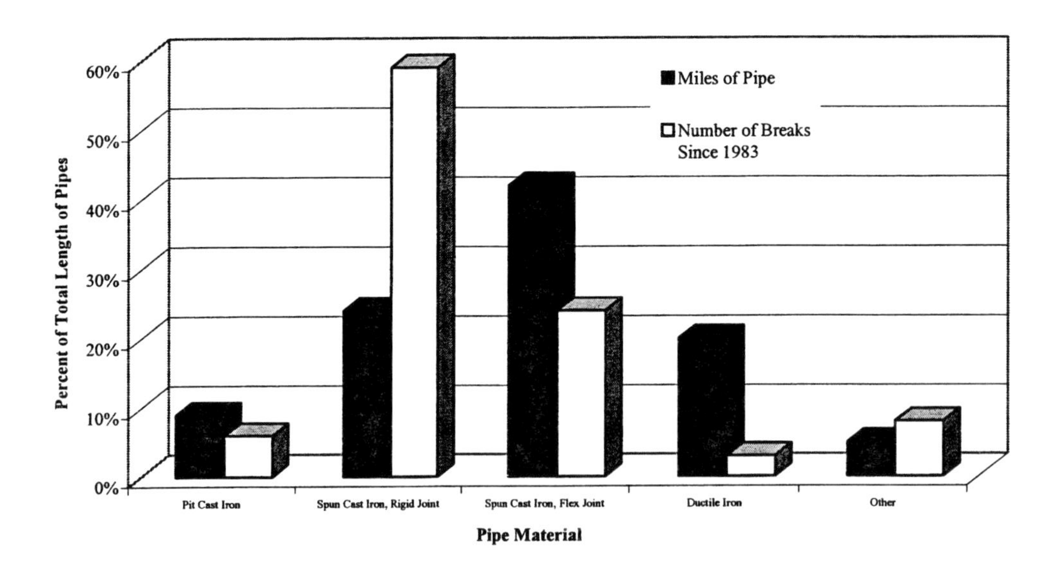

Figure 4.1 Example of material analysis

The data required to conduct these types of spatial analyses are fairly basic. Obviously, the utility must maintain a record of water main break occurrences. Associated with each break record must be some type of spatial locator. Typically, this would be a street address. A GIS could then reference the addresses and relate main breaks to specific locations in the distribution system. For utilities without a GIS, main breaks could be plotted manually on maps using the street address so that geographic patterns could be observed.

Figure 4.2 presents a simple example of using a GIS to evaluate spatial relationship between main breaks. This figure shows a portion of a distribution system, with pipes coded to indicate the number of main breaks that have occurred.

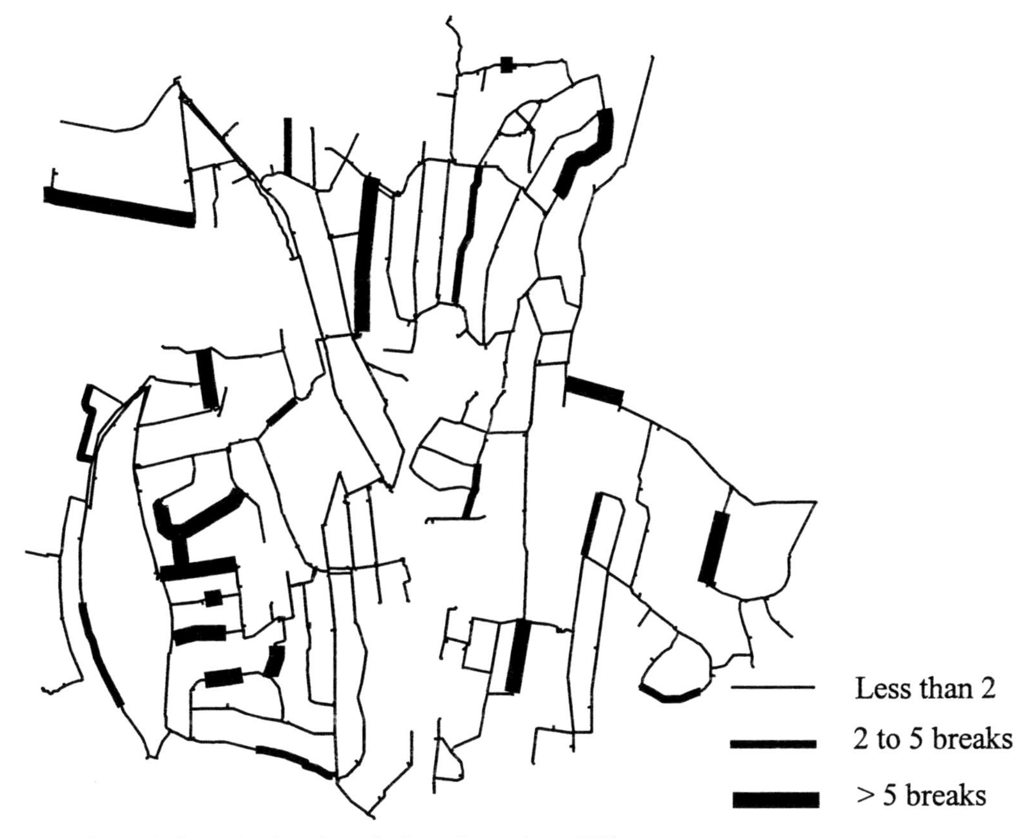

Figure 4.2 Example of spatial analysis of main breaks using GIS

TEMPORAL ANALYSIS

A review of water main breaks over time is another simple analysis that can be performed if data are collected and maintained over an extended period. Again, the only data needed are a record of main breaks and the date on which they occurred. The number of water main breaks occurring in a given year is impacted by numerous factors, with weather being one of the chief factors. Figure 4.3 illustrates an example of analyzing water main breaks over time. The plot shows that the total number of breaks generally increased over time. This is true even if one accounts for years in which the number of main breaks was particularly high due to weather conditions.

Another way to examine water main breaks over time is illustrated in Figure 4.4 for the same utility as in Figure 4.2. Here the break rate, expressed in terms of breaks per 100 miles per year, is considered instead of the total number of breaks. This type of analysis is useful to

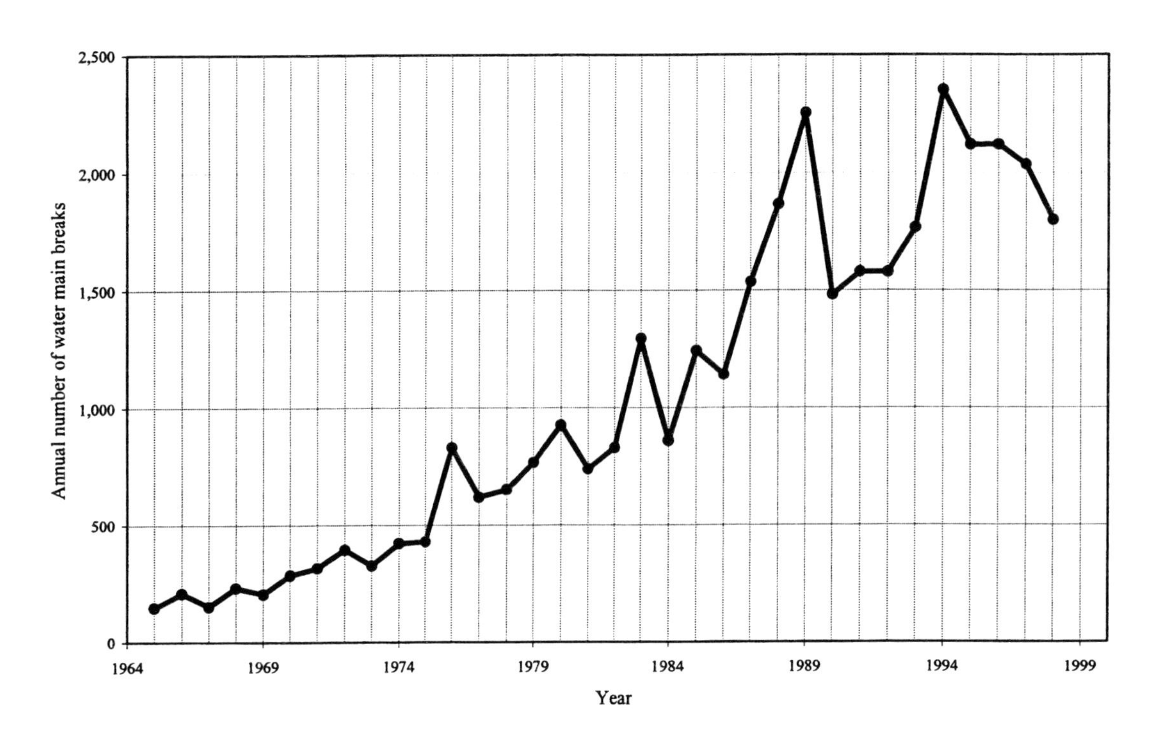

Figure 4.3 Number of water main breaks over time

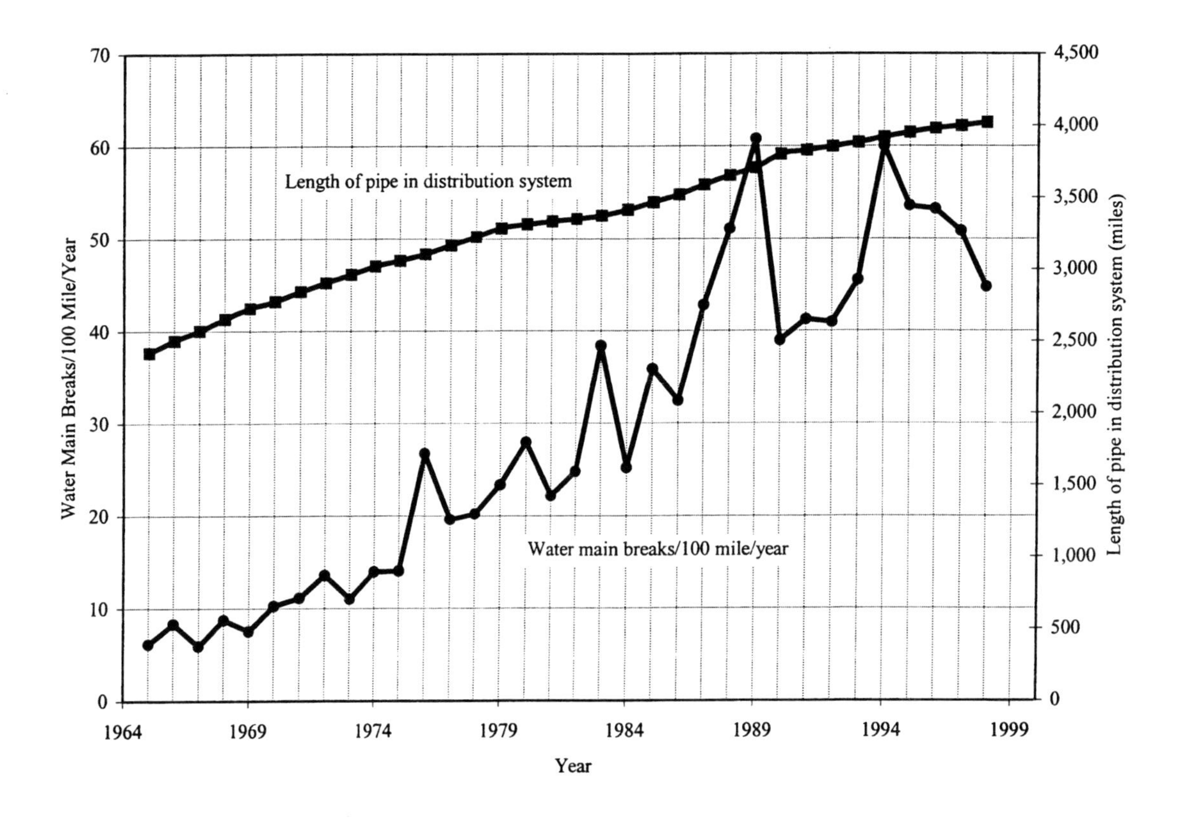

Figure 4.4 Water main break rate over time

account for a growing system that may be experiencing more water main breaks annually simply because the amount of pipe in the system has grown rapidly. In Figure 4.4 the break rate has also grown over time, indicating that the frequency of breaks is increasing despite a corresponding increase in length of pipe in the system.

BREAK FORECASTING

Another use of water main break data is to predict future main breaks. As described in Chapter 2, various attempts have been made to forecast water main breaks based upon historic data. In simplest form, the utility would examine the annual number of main breaks in the past and forecast the number into the future based upon the historic trend. A more detailed analysis would attempt to forecast the occurrence of breaks on individual pipes.

One basic use of these forecasts is to plan manpower and budgetary requirements for upcoming years. The more detailed predictions of future breaks on individual pipes should not be used implicitly. Water main breaks are the result of a variety of environmental conditions, none of which can be predicted with any certainty. Rather, predictions of main breaks for individual pipes can be used for comparison purposes to prioritize pipe renewal needs. That is, pipes with greater predicted frequencies of breaks are more likely to have a higher renewal priority than pipes with lower frequencies of predicted breaks.

The data needed to develop predictions of future main breaks ranges from simple to complex. For example, a forecast of annual number of main breaks could be developed only using the number of breaks occurring in previous years. Figure 4.5 uses the same break history shown in Figure 4.4. The anticipated number of future breaks per year could be projected using simple linear or exponential regressions of the historic data. As shown, the two regressions yield considerably different results. The linear regression projects main breaks to slightly exceed 2,500 per year by 2004. The exponential regression, which actually fits the data better as indicated by the higher R-squared value, projects more than 4,500 main breaks per year by 2004.

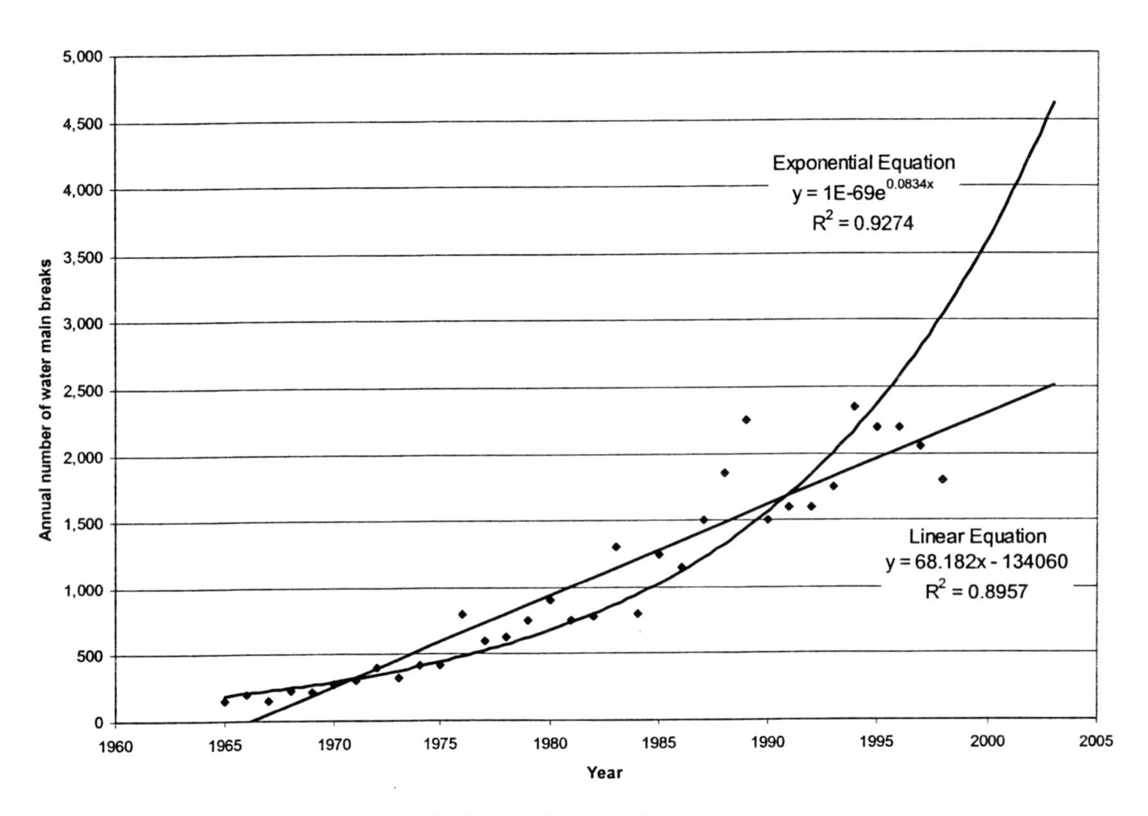

Figure 4.5 Example of main break forecasting

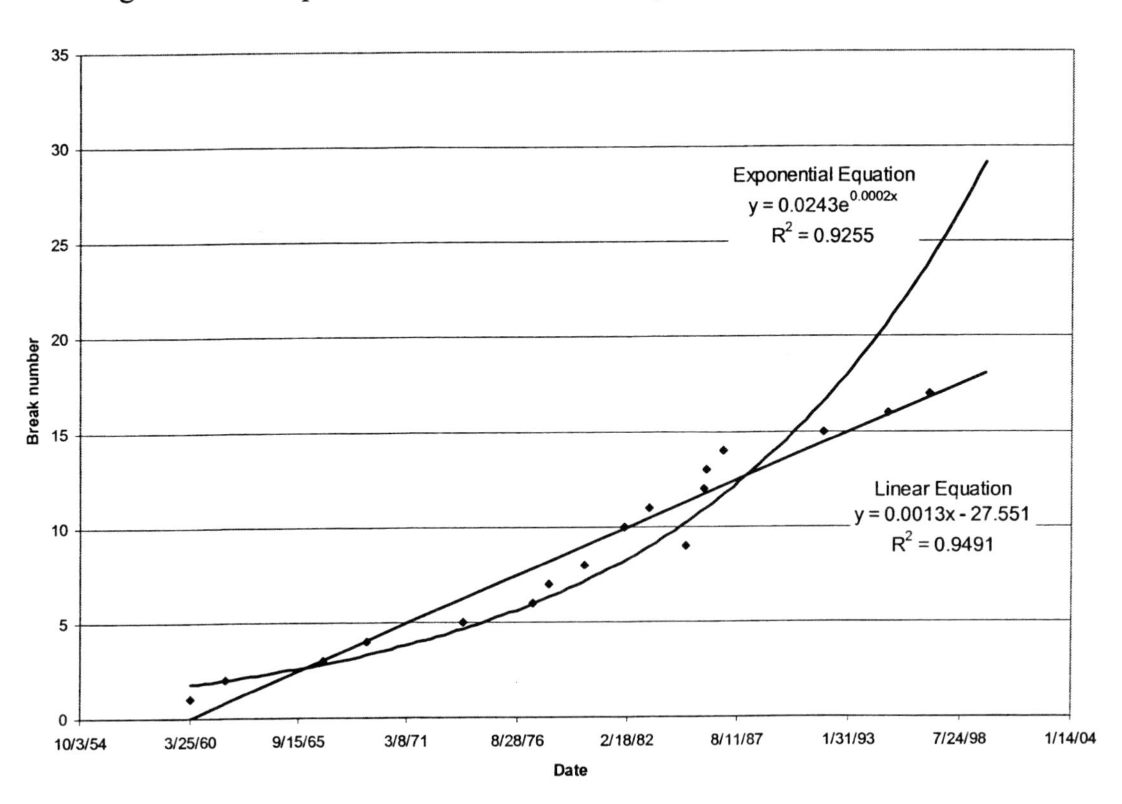

Figure 4.6 Example of main break forecasting for individual pipe

A more detailed analysis could examine the break histories and attempt to predict future breaks for individual pipes. Figure 4.6 uses the example of Pipe ID 1113 from Table 4.1, and shows similar regression analysis as Figure 4.5. Forecasts of future breaks produced in this manner should not be considered predictors of actual break events. Numerous factors impact the occurrence of main breaks, and these factors are too variable to allow for detailed and accurate predictions. However, predictions made in this manner are useful and appropriate for certain purposes. The system-wide forecasting of breaks can be a useful tool for planning annual manpower and budgetary requirements. Break predictions for individual pipes, while not to be taken literally, provide a method for comparing the probability of failure among the pipes in the system. This can be a valuable first step in developing a prioritized renewal program.

ECONOMIC ANALYSIS

A pipe is an asset of a water utility. The total life cycle cost of the pipe includes its installation cost and maintenance costs. For a pipe, the only maintenance costs is typically the cost of repairing main breaks. If sufficient and adequately detailed water main break data are available, an economic model can be developed that considers the cumulative costs of future main breaks on a pipe, and the ultimate cost of replacing that pipe. The results of the model can be used to help prioritize water mains for renewal. An example is provided here from an economic modeling study performed for the SLCWC (Grablutz and Hanneken 2000).

The first step in such economic modeling is to predict future water main breaks as described previously. The break forecasting module relies on historic break data to predict future breaks for individual pipes. The SLCWC Leak Database provided the data used in this modeling. An Excel spreadsheet was developed to process the data by eliminating irrelevant records and identifying inconsistent records requiring further research. The spreadsheet then assigned a Pipe ID to each record, and generated a file used as input for the next step. The Pipe ID was a unique identifier for an individual pipe, incorporating the installation year, material, and diameter of pipe in order to avoid duplicate records.

Regression analysis was used to predict future breaks on a particular pipe. At first it was assumed that the break trends would follow an exponential pattern. However, it was determined that in many cases a linear function better described the pattern of water main breaks for a given pipe. Therefore, the program performed linear and exponential regression analyses for the breaks for each Pipe ID. In order to conduct a representative regression analysis, at least three breaks must have occurred on the pipe. The model computed R-squared values for the linear and exponential regressions. The regression equation with the higher R-squared value best was assumed to represent the water main break trend for that particular pipe. Thus, either the linear or exponential equation was used to predict future water main breaks for a pipe.

The second component of the economic model was the break-even analysis. This analysis computed the total present worth cost associated with the future life of a pipe. The costs associated with the future life of the pipe included repair costs when it fails, and ultimately, replacement of the pipe. The financially optimum time to replace the pipe is when this total present worth cost is at a minimum. Figure 4.7 illustrates the fundamentals of the approach graphically. The present worth costs associated with the Repair Cost curve increase over time since this is a cumulative curve. The replacement cost for the pipe was assumed to stay the same over time (i.e., no inflation), and therefore, the present worth cost of replacement decrease over time. The Total Cost Curve is the sum of the Repair Cost Curve and the Replacement Cost curve. The economically optimum time to replace the pipe is when the Total Cost Curve is at a minimum, or in the example of Figure 4.7, early in 2007.

In conducting the break-even analysis, it is important to consider all of the costs associated with a water main break. Typically, a water utility will only consider the so-called direct costs, including labor, materials, roadway repair, damage claims, etc. However, the true costs of a water main break should include indirect costs such as the value of lost water, traffic disruption, service interruption, etc. Although it is difficult to assign a dollar value to these indirect costs, they should be considered if a true impact of main breaks on customers is to be evaluated. In order to account for these indirect costs, the SLCWC model provides the user with the ability to assign these indirect costs a value represented as a percentage of the direct costs. In the study, indirect costs were estimated to be between 20% and 40% of the direct cost of a water main break.

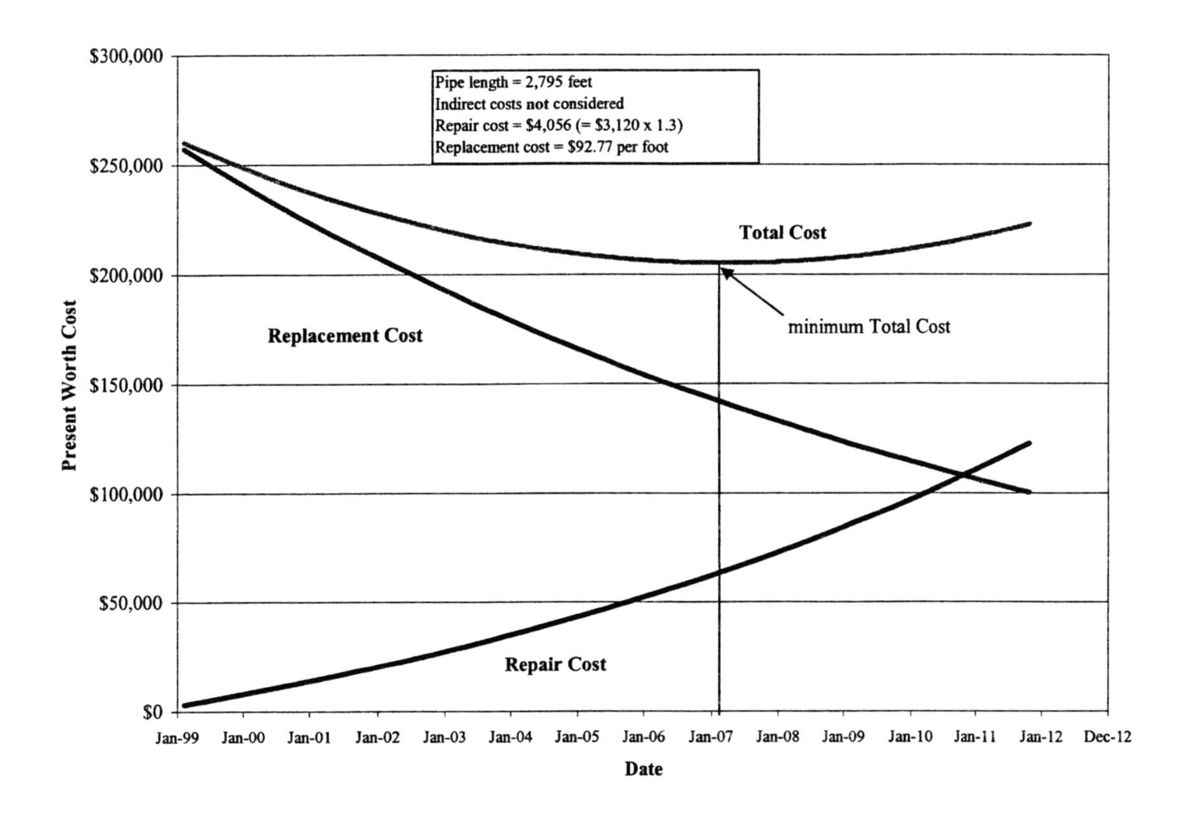

Figure 4.7 Example of economic break-even analysis (no indirect costs)

In the example in Figure 4.7, indirect costs were not considered. Figure 4.8 illustrates the impact of indirect costs on the economically optimum replacement time. This break-even analysis is for the same pipe shown in Figure 4.7, except that in Figure 4.8 indirect costs were assumed to be 30% of the direct costs. The impact of the indirect costs is to change the financially optimum replacement time from year 2007 to 2005.

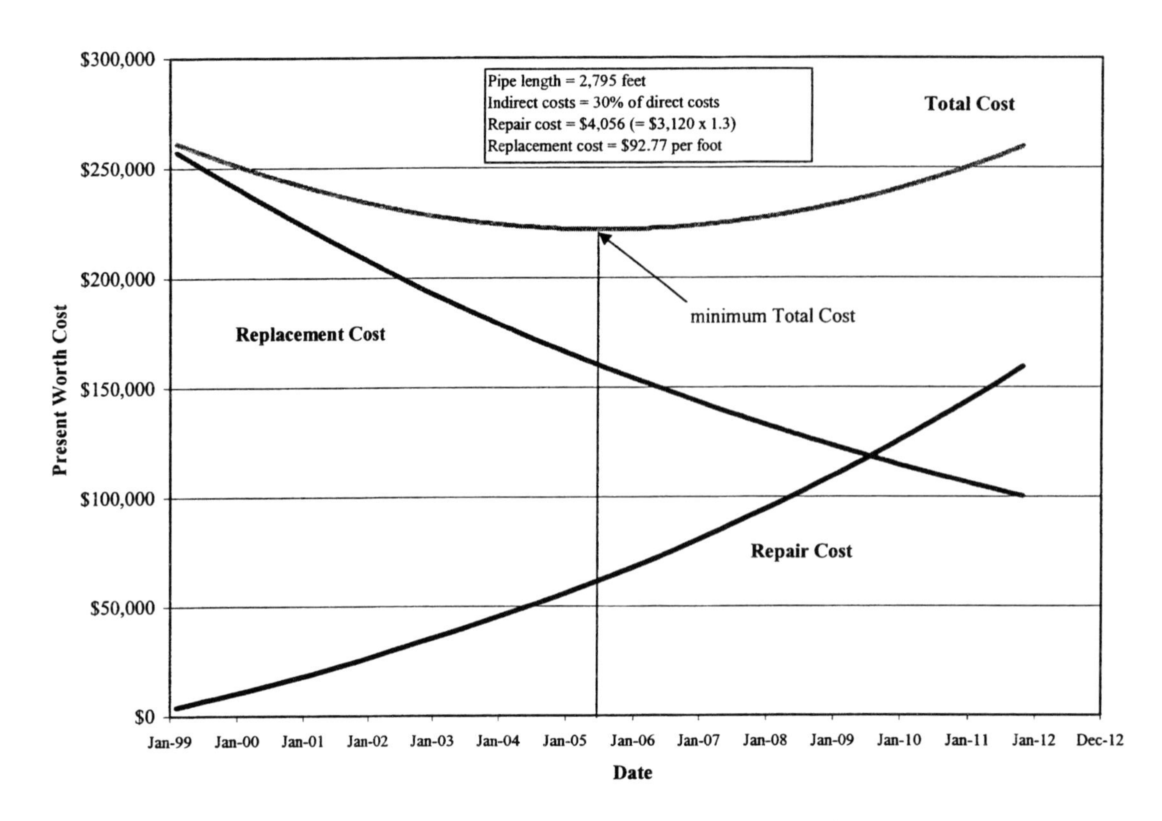

Figure 4.8 Example of economic break-even analysis (indirect costs included)

MECHANISTIC MODELING

The mechanistic model of water main breaks developed for this project involved the detailed estimation of the growth of corrosion pits on cast iron pipe, the resultant loss of wall thickness, and the reduction in pipe strength over time. The residual strength of the pipe is compared against estimates of various loads and stresses on the pipe to assess the pipe's vulnerability to breaking. Mechanistic modeling is a more complex analysis method than the analysis of main break occurrence, material, spatial or temporal analysis, break forecasting, or economic analysis. Because of the large number of factors that must be considered in order to perform a mechanistic modeling analysis of water mains, this technique has the most uncertainty. This limitation is offset, however, by the ability to assess and compare all water mains in a distribution system using consistent objective methodologies, and to develop a prioritized list of water mains according to their vulnerability to failure. The mechanistic model can also be applied to water mains that have not yet broken. The other planning techniques described previously can only be applied to water mains with a break history. Thus, the mechanistic modeling approach represents the most proactive approach to distribution system renewal.

The residual life of a pipe is the crucial factor in determining water main renewal priorities. If it was possible to identify pipes that were vulnerable to failure, then a prioritized replacement program could be initiated focusing on these vulnerable pipes. A pipe fails when the loads exerted on it create stresses that exceed the residual strength of the pipe.

Pipes are manufactured to withstand certain loads described in terms of the longitudinal tensile strength, flexural strength to withstand bending as a beam, ring strength to withstand crushing load given by modulus of rupture, and bursting strength to withstand radial pressure (see Figures 4.9 and 4.10). These loads are the failure-causing loads, and a pipe is expected to fail when it encounters such a load. In the field, a buried water main has to withstand the earth load, truck/live load, working water pressure, and water hammer pressure in addition to less frequent loads from frost, expansive soil, and stresses on the pipe wall due to thermal expansion and contraction. Typically, the ratio between the strength of the pipe and the stresses on a pipe is thought of as a margin of safety or safety factor (SF). The pipe material and thickness are designed and selected to meet a certain SF. Once a pipe is put into use, it faces a deterioration process and continuously loses wall thickness. The SF of the pipe decreases as the residual strength of the pipe decreases along with pipe wall thickness.

It is worthwhile to discuss the concept and use of SF in detail given the important role they play in the mechanistic model. SF are used as indicators of the strength of a pipe in relation to the maximum stresses to which the pipe could reasonably be subjected. Specifically, SF is the ratio of the remaining strength, in psi, to the maximum reasonable stress on the pipe. Separate SFs are calculated for stresses that act in different directions. SFs are calculated for longitudinal (tensile) stress, hoop (bursting tensile) stress, ring stress, and bending (flexural or beam) stress. Since hoop stress and ring stress are additive at some locations in the pipe wall, a combined SF for hoop and ring stress is calculated. Similarly, a SF for combined longitudinal stress and bending stress is calculated.

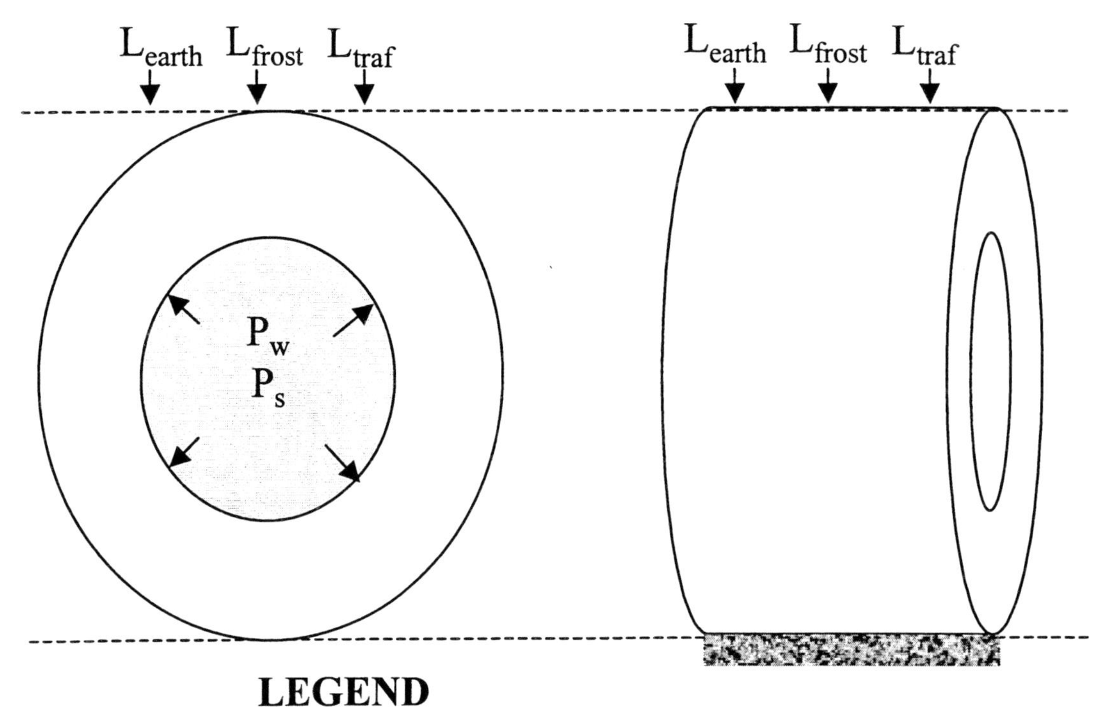

LEGEND

L_{earth}=Earth Load
L_{frost}= Frost Load
$L_{traffic}$= Traffic Load

P_w= Working Pressure

P_s= Surge Pressure

Figure 4.9 Illustration of external and internal loads on a buried pipe

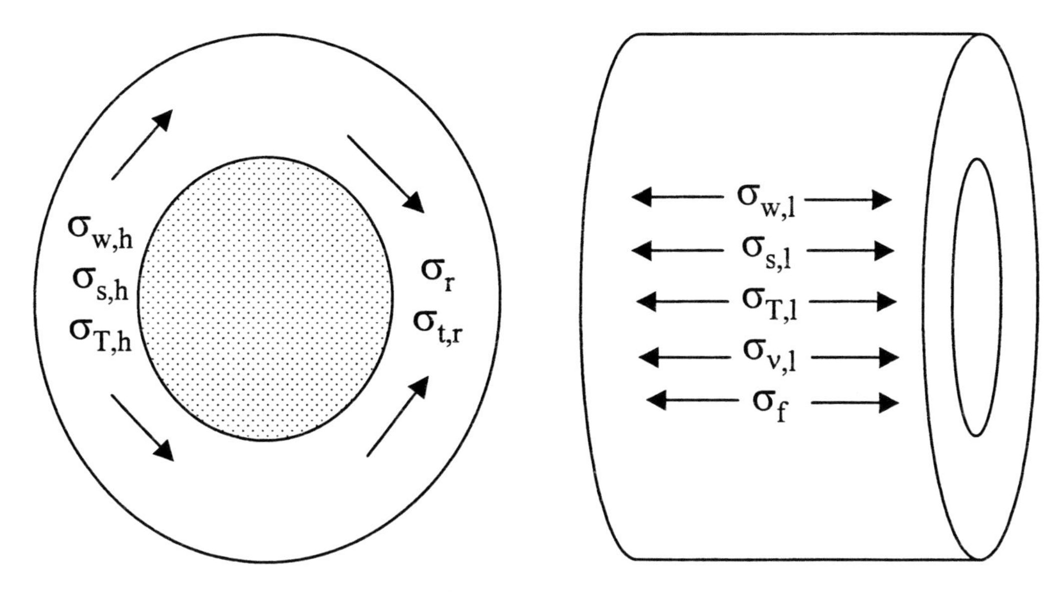

LEGEND

$\sigma_{w,h}$ = Hoop Stress (Working Pressure)
$\sigma_{s,h}$ = Hoop Stress (Surge Pressure)
σ_r = Ring Stress (External Loads)
$\sigma_{t,r}$ = Ring Stress (Thermally Induced)
$\sigma_{w,l}$ = Longitudinal Stress (Working Pressure)
$\sigma_{s,l}$ = Longitudinal Stress (Surge Pressure)
$\sigma_{T,l}$ = Longitudinal Stress (Thermally Induced)
$\sigma_{v,l}$ = Longitudinal Stress due to Poisson Effect
σ_f = Flexural Stress

Figure 4.10 Illustration of stresses on buried pipe

Since the occurrence of large loads on a water main, such as frost load and water hammer, occur at irregular and generally unpredictable intervals, SF represent a "worst case" estimation of the remaining strength of the a pipe compared to the maximum stresses to which it can be subjected. Theoretically, the SF of a pipe will be below 1.0 at the time of failure (i.e., the stresses on the pipe exceed the remaining strength).

When the mechanistic model is applied to a database of pipes that have broken, there is a range of SF predicted, some greater than 1.0 and others less than 1.0. When the model estimates the SF of a pipe to be greater than 1.0 at the time it failed, this implies that either the maximum loads to which the pipe was subjected was under-estimated or the remaining strength of the pipe were over-estimated.

The critical (minimum) SF predicted by the model should also be consistent with the mode of failure. Pipes with longitudinal failures (splits) would be expected to have low SF for combined ring and hoop stress. Pipes experiencing circumferential breaks would be expected to have low SF for combined longitudinal and bending (flexural) stress. Pipes with pinhole leaks should show low SF due to small remaining wall thickness. Conceptually a pipe with a pinhole leak could fail with a circumferential break if subject to excessive forces.

Pipes located in regions prone to freezing temperatures sometimes experience an additional load (frost load) caused by frost heaving of surrounding soil. Similarly, wide and rapid temperature variations in the soil-pipe-water environment lead to additional thermal stresses on the pipe. Leakage in pipes and bad construction practices around the pipe sometimes lead to the pipe bed disruption thereby making it prone to breakage due to beam action. In addition to the increased loads on the pipe, the pipe's structural integrity is deteriorated by corrosion at a rate dependent on the pipe material type, protection type, characteristics of the surrounding soil, and the hydraulic and chemical properties of the water flowing in the pipe. Corrosive soils accelerate the development of corrosion pits on the pipe outside surface. Corrosive waters flowing in the pipe also accelerate the graphitization and the eventual reduction in pipe wall thickness. The quality of water inside an unlined pipe also dictates the growth of tubercles within the pipe that tend to reduce the effective diameter of the pipe as well as increase the pipe roughness. To compensate for the increased head loss, the excessive use of a pump may be needed to maintain the working pressure. This can cause water hammer conditions and result in additional stress on the weakened pipe. The mechanistic model developed in this project can be used to assess the residual strength of a pipe and compare it to the estimated stresses experienced by the pipe. The ratio of residual strength to working stress is considered the SF. The SF can then be used to prioritize pipes for replacement.

This section describes the development of the mechanistic model as a tool for prioritizing cast iron pipes for renewal. The data requirements for the model are described along with a description of how to appropriately apply the model given its features and limitations.

Objectives of Modeling

The objectives of the models developed in this project were to:

- Estimate the residual strength of cast iron water mains,
- Calculate maximum loads to which the water mains are exposed,
- Calculate a SF for each pipe as the residual strength of the pipe divided by the pipe stresses resulting from the maximum loads to which the pipe is subjected, and
- Prioritize the water mains for replacement based upon the calculated SF for individual pipes.

The mechanistic model, as it is presented here, is limited to cast iron water mains subject to external corrosion from corrosive soils. The model does not consider external corrosion due to the effects of stray direct currents. The model has the provision to account for internal corrosion through a user-specified "corrosion rate" parameter. However, the model itself cannot predict internal corrosion rates.

Modeling Approach

A mechanistic modeling technique was developed and implemented in a spreadsheet application. The mechanistic model is made up of four modules: (1) the Pipe Load Module (PLM), (2) Pipe Deterioration Module (PDM), (3) Statistical Correlation Module (SCM), and (4) Pipe Break Module (PBM). The functions of the four modules are described below.

Pipe Load Module (PLM)

The function of the PLM is to estimate the maximum probable loads to which the pipe may be exposed and the resulting stresses that would be created within the pipe. The PLM considers both internal and external loads. External loads considered in the model include earth load, traffic load, and frost load where appropriate. Internal loads considered by the PLM include internal working pressure and surge pressures that may occur. The PLM also considers thermally-induced stresses on the pipe.

Four separate component5s of stresses on the pipe are analyzed by the PLM as shown in Table 4.2.

Hoop stress and ring stress result in longitudinal breaks while flexural stresses and longitudinal stresses result in circumferential breaks. Appendix C contains a detailed presentation of the equations used by the four modules, their basis, and data requirements.

Pipe Deterioration Module (PDM)

The function of the PDM is to estimate the depth of external corrosion pits on cast iron water mains. The corrosion rate is based upon a relationship described by Rossum (1969). The Rossum relationship describes the depth of the corrosion pit to be a function of soil characteristics (pH, resistivity, and soil aeration) and pipe characteristics and pipe surface area exposed to the soil.

Table 4.2

Four components of stresses analyzed by PLM

Stress	Resulting from
Hoop stress	Internal working pressure and surge pressure
Ring stress	External crushing loads on pipe with bedding support
Flexural (bending) stress	Loss of bedding support below a portion of the pipe under external loads
Longitudinal stress	Working pressure, surge pressure, and thermal stresses

Statistical Correlation Module (SCM)

The SCM is used to calculate the residual strength of the pipe as a function of remaining pipe wall thickness (i.e., original wall thickness minus corrosion pit depth). The calculation of residual strength is based upon the original strength and thickness of the pipe and a statistical correlation of laboratory data relating pipe wall thickness to modulus of rupture, tensile strength, ring modulus, and fracture toughness.

Based upon a review of material laboratory testing data, it appears that there is much less uncertainty in the prediction of pipe residual strength than there is in the prediction of loads and resultant stresses on pipes.

Pipe Break Module (PBM)

The PBM calculates "safety factors" (SF) as the ratio of residual pipe strength to maximum stress on the pipe. Theoretically, the pipe will fail when the SF falls below a value of 1.0 (i.e., the stresses on the pipe exceed the residual strength). Six SF are calculated by the PBM as follows.

- Ring stress SF
- Hoop stress SF
- Tensile (longitudinal) stress SF
- Bending stress SF
- Combined ring and hoop stress SF
- Combined tensile and bending stress SF

The PBM compares the six SF and finds the minimum value for every pipe in a database of cast iron pipes. The pipes are then sorted by increasing SF. The pipes with the lowest SF are assumed to be more vulnerable to failure, and thus should be targeted for replacement sooner than those pipes with higher SF. Thus, the final product of the PBM is a prioritized replacement list of cast iron water mains.

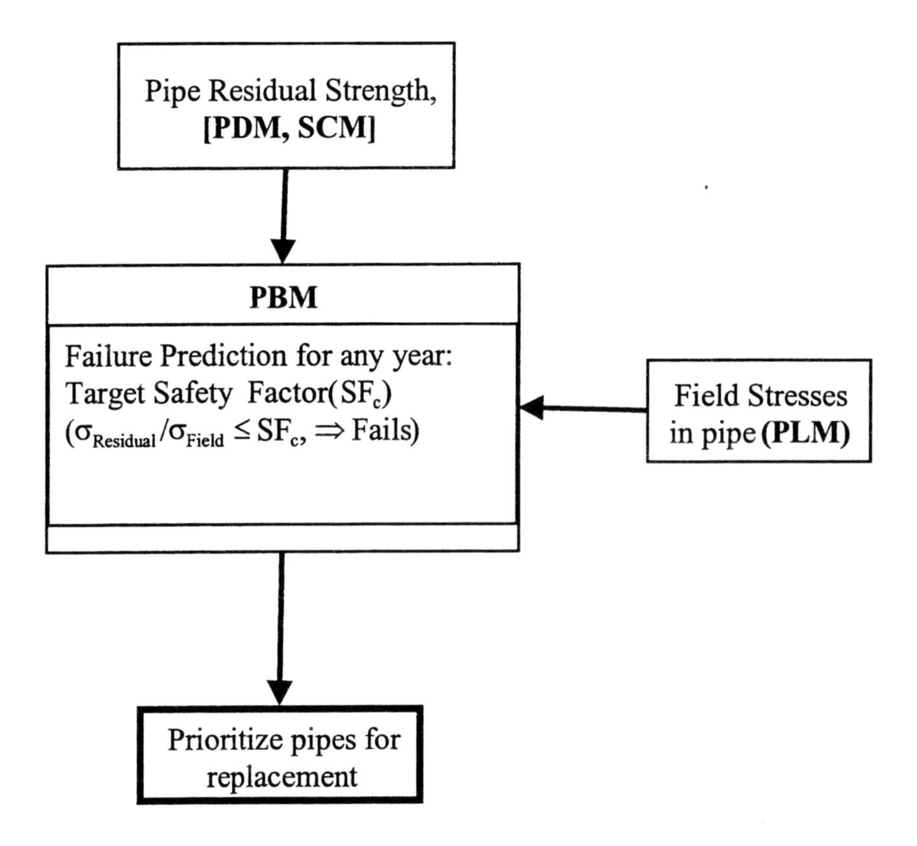

Figure 4.11 Flow chart of pipe failure analysis

The diagram in Figure 4.11 illustrates the interactions between the modules of the modeling system to determine which pipes need to be replaced and the order they should be replaced.

Using the Model

Because the processes simulated by the model are complex and highly variable the model cannot be expected to predict the month or even the year of a water main break. The loads to which pipes are subjected are highly variable in their magnitude and frequency of occurrence. Similarly, the corrosion processes affecting cast iron pipe are neither uniform nor easy to predict. However, the relationship between wall thickness and residual strength of cast iron pipes is moderately well established and predictable.

Therefore, the mechanistic model is more appropriately used as a prioritization tool rather than an absolute predictor of water main breaks. The SF estimated by the model is an index of the vulnerability of the pipe to failure. These indices will be more accurate when data collected from main break repairs, primarily wall thickness, are used to adjust model predictions of corrosion rate.

The mechanistic model was developed specifically for cast iron pipe. It addresses the loss of pipe strength due to external corrosion, and does not attempt to estimate the impact of internal corrosion. The user can, however, input an overall corrosion rate to address both forms of corrosion if suitable historic data are available. The model only considers typical loads on a pipe, including earth loads, traffic loads, frost loads, etc. Appendix C provides additional detail on the loads considered and how the model computes them. The model does not account for potentially significant loads resulting from earthquakes, landslides, differential ground movement due to moisture gradients, or disruptions due to nearby excavation. Although these loads can be considerable and lead to main breaks, they cannot be accounted for in a model aimed at "typical" conditions. The impact of these loads should certainly be considered separately by a utility planning a water main renewal program.

The steps by which a water utility would apply the model are as follows:

1. The user selects a cast iron water main or several mains from a break database. The database must contain the data required by the mechanistic model including date of installation and the date of break.
2. When necessary, divide the pipes into smaller sections or segments with unique soil and environmental characteristics. The analysis procedure requires that any pipe being analyzed must be in a uniform environment. For example, a pipe being analyzed cannot have parts both in corrosive and non-corrosive soils. In order to avoid such a situation, the selected pipe will be divided into sections based on what types of soils they pass through. A generalized approach currently used by some utilities is to divide the area in which the water distribution system exists into grids. Each grid has its own soil and pertinent properties assigned to it. The portion of each pipe passing through a grid will be referred to as a section of that pipe.
3. Specify a target SF for analysis (e.g., 1.2, 1.5, etc.). The user needs to specify desired target SF as a criterion for renewal. This value is used to determine when a particular pipe section is near failure.

4. Pipe break analysis then proceeds as follows:

<u>Step i:</u> Determine the following external and internal loads on pipe from PLM:

   a. Earth load
   b. Traffic load
   c. Frost load, if applicable
   d. Internal pressure (working water pressure)
   e. Internal surge pressure (water hammer)

<u>Step ii:</u> Determine the following stresses on pipe from external and internal loads using PLM:

   a. Ring stress from ring crushing loads (frost, expansive soil, earth, and traffic)
   b. Longitudinal stress from internal pressures
   c. Hoop stress from internal pressures
   d. Thermal stresses (longitudinal and ring)
   e. Flexural stress from beam action from external loads (frost, expansive soil, earth, and traffic)

<u>Step iii:</u> Estimate the following structural conditions of the pipe using PDM:

   a. Wall thickness reduction due to corrosion of the pipe over time
   b. Stress concentration and susceptibility to fracture due to pit growth

<u>Step iv:</u> Calculate the reduction in residual strength due to corrosion pit growth using SCM.

   a. Reduction in modulus of rupture (i.e., residual strength) for both beam action and ring deflection
   b. Confirm prediction of pit depth and residual strength with materials testing lab data if available.

Step v: Compute residual SF by comparing loads on pipe with the pipe's structural strength condition. This will be accomplished using PBM. The SF will be computed and evaluated for the following stress components:

a. Hoop stress
b. Ring stress
c. Longitudinal stress
d. Flexural (beam action) stress
e. Combined hoop and ring stress
f. Combined flexural and longitudinal stress

Step vi: The user will calibrate the model results by adjusting the input data to the model as follows:

a. Adjust corrosion rate, beam span, and other input parameters to achieve target SF for pipes that have already failed.
b. Confirm target SF for pipe failure and failure mode (i.e., longitudinal break (hoop/ring SF) or circumferential break (flexural/longitudinal SF).

Step vii: Compare residual SF with target SF. This results in a list of the pipes where the calculated SF is less than the user-specified target SF. The predicted failure mode of the pipe, such as beam stress or ring failure, is predicted by the stress component with the lowest SF.

5. Develop prioritized renewal program

   a. Apply model to pipe inventory database (cast iron pipes only) to calculate SF
   b. List pipes with computed SF less than the target SF
   c. Order pipes in increasing order of SF. This is a prioritized list of cast iron water mains for replacement prioritized by their potential for failure.

CHAPTER 5

MODELING CASE STUDY

The mechanistic model described in Chapter 4 was tested using actual data to evaluate its applicability for use by water utilities. The RMOC water main break database was used to test the model. The following sections describe the modeling process and the final results.

INPUT DATA

The various data inputs described in Chapter 4 were gathered from both the RMOC water main inventory and the RMOC main break database. Assumptions were made when the required data were not available. Pipes were selected from the break database for analysis as long as the pipe was also represented in the RMOC water main inventory (so that background data about the pipe were available) and according to the following criteria:

- material type was cast iron (pit or spun cast)
- year installed was provided
- soil class was provided
- diameter less than 310 mm (~12 in)

A total of 4,268 break records met these criteria and were used in the analysis. The data fields extracted from the water main inventory were:

- pipeid
- size (diameter)
- length
- soil class
- material type
- year installed

The pipeid field provides a unique identifier for the pipe, and also provides the link to data in the main break database. The units of the pipe size were converted from millimeters to inches and the lengths were converted from meters to feet. The soil class is a designation used by RMOC to describe the soil environment of the pipe. Only cast iron pipes from the database were used for the analysis. These are designated in the database as CI (cement lined cast iron) and UCI (unlined cast iron). Pipes installed prior to and in 1929 were assumed to be pit cast and those after 1929 as spun cast. Since the inventory records had no associated pipe depth of cover, the depth of cover was assumed to be the average of the depth obtained from the break database [6.56ft (2m)]. The "AnalysisYear" was set to the year in which each pipe failed. In other words, the model simulates the aging of all pipes from the time of installation year until it fails. The "TrafficType" for all the pipes was assumed to be heavy and the "PavementType" as paved. "WorkingPressure" was assumed to be 100 psi. The "BeamSpan" was assumed to be 4 ft for the purposes of calculating bending stress on the pipes.

Table 5.1 summarizes the various data inputs and from where the data were obtained. Table 5.2 provides the assumed values used where required data were not available. Table 5.3 lists the soil properties associated with each RMOC soil class. The soil types 1 through 5 show a progressive increase in corrosivity, with soil type 5 (organic clays) being the most corrosive. For soil type 8 (unclassified) the properties of soil type 5 were assigned. Note that although category 8 is "Unclassified", it was assigned fairly corrosive soil characteristics to represent a worst case deterioration scenario.

Table 5.1

Model data inputs for the RMOC test case

Input	Purpose/description	Source
Pipe ID	Identify individual pipes	Main break database
Region ID	Characterize certain weather related characteristics	Characteristics were assumed (see Table 5.2)
Soil class	Assign corrosion characteristics for the soil surrounding the pipe	Main break database
Analysis year	The year in which the main break occurred representing the end point of the pipe deterioration	Main break database
Pipe type	The material of the pipe that failed	Main break database (either CI for cement lined cast iron or UCI for unlined cast iron)
Pipe diameter	The diameter of the pipe (in inches)	Main break database
Pipe year installed	The installation year of the pipe	Main break database
Pipe length	The length of the pipe in feet	Main break database
Traffic type	Assign live loading characteristics	Assumed (see Table 5.2)
Working pressure	Compute loads due to water pressure	Main break database
Pipe depth	Compute soil load on pipe	Water main inventory
Pavement type	Compute dead load on pipe	Water main inventory
Beam span	Compute bending load on pipe	Assumed (see Table 5.2)

Table 5.2

Assumptions used in RMOC test case of the model

Parameter	Assumed input value
Region ID characteristics	
Minimum yearly temperature	-20 °F
Maximum yearly temperature	85 °F
Maximum frost depth	50 in
Maximum sudden water temperature change	6 °F
Maximum freeze days	60 days
Minimum water temperature	32 °F
Maximum water temperature	60 °F
Water velocity change	2 ft/s
Soil characteristics	See Table 5.3
Traffic type	heavy traffic
Beam span	4 ft

Table 5.3

RMOC modeling test case soil characteristic assumptions

Soil class	Soil description	Soil pH	Soil resistivity ohm-cm	Soil aeration	Soil moisture %	Soil liquid limit	Soil density lb/ft^3	Soil porosity %	Soil expansive	Frost susceptible
1	Coarse: gravels, sands, loamy sands	8	1500	good	15	21	135	35	n	n
2	Moderately coarse: sandy loam, loam, silt loam	7	1200	good	20	32	130	45	n	n
3	Moderately fine: sandy clay loam, silt, silty clay loam, clay loam	6	1000	fair	20	32	120	45	y	y
4	Fine: sandy clay, silty clay, clay, heavy clay	6	1000	fair	20	26	120	40	y	y
5	Organic	5	700	poor	25	59	100	60	y	y
7	Bedrock	8	2000	good	15	39	135	50	n	n
8	Unclassified	5	700	poor	15	21	100	35	y	y

MODEL RESULTS

The results of applying the model to the RMOC main break database and water main inventory were evaluated in order to determine the effectiveness and applicability of the model for use by utilities in prioritizing main replacement programs. In evaluating the model, the SF for the combination of hoop and ring stresses (SF_{h-r}) and flexural and longitudinal stresses (SF_{f-l}) were considered. These SF represent the worst case combination of stresses and strengths for the pipe, and are, thus, most appropriate for analyzing past failures and probabilities of future failures.

The average SF for the 4,268 pipes in the break inventory was 1.64 for SF_{f-l} with a standard deviation of 0.49. The average SF_{h-r} was also 1.64 with a standard deviation of 0.97. To test the sensitivity of input parameters, a second model runs was made. In this second model run the beam span (i.e., the unsupported length of pipe) was changed from 4 ft to 8 ft. This resulted in average SF_{f-l} and SF_{h-r} of 0.80 and 1.51, respectively. The corresponding standard deviations were 0.47 and 0.97. The significant decrease in SF_{f-l} highlights the importance of beam span in determining main breaks.

It was expected that the SF would vary based upon soil category. More corrosive soils should have generally lower SF than non-corrosive soils. Figure 5.1 shows the average SF_{f-l} and SF_{h-r} computed by the model for pipes within each soil category. As expected, less corrosive soils (categories 1, 2, 4, and 7) have generally higher SF than do the corrosive soils (categories 5 and 8).

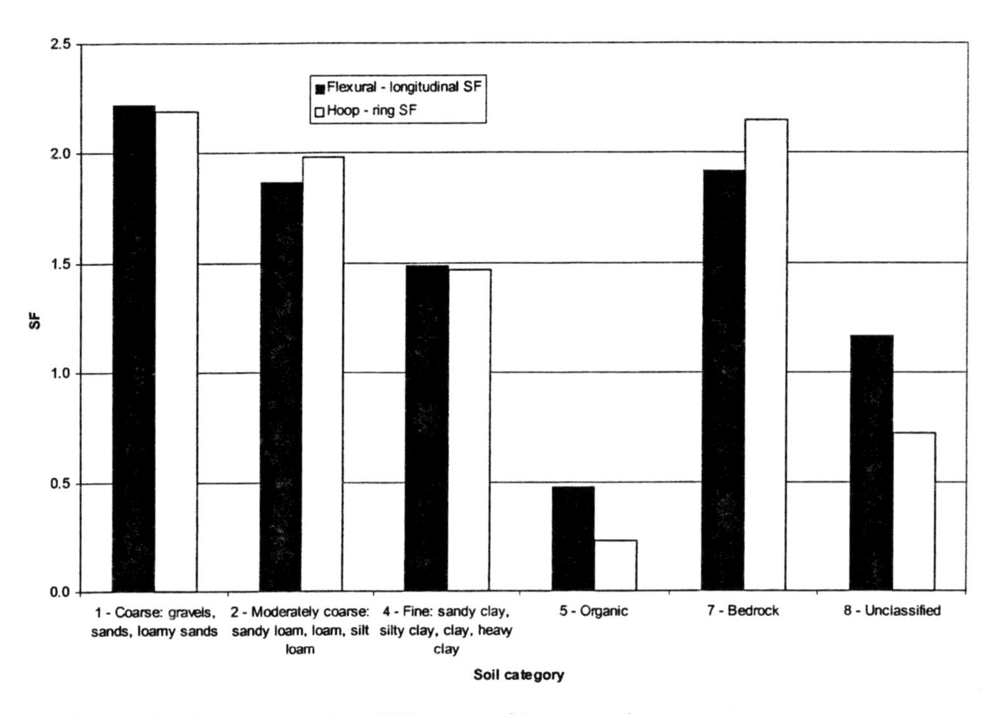

Figure 5.1 SF for RMOC test case for different soil categories

The computed SF were also examined according to pipe diameter. Figure 5.2 shows the computed SF for the five pipe diameters in the break database. The SF_{f-l} were fairly consistent across all the diameters, ranging from a low of 1.44 for the 4-inch pipe to a high of 1.79 for the 12-inch pipe. On the other hand, the SF_{h-r} showed considerable variation over the different diameters. The SF_{h-r} for the 4-inch pipe was 3.77, considerably above the theoretical SF at which a pipe would fail (1.0). This would indicate that the failure mode for 4-inch pipes is most likely due to flexural-longitudinal stresses. The SF_{h-r} for 10-inch and 12-inch pipe were both below 1.0 at 0.72 and 0.86, respectively. This would indicate that the failure mode for these pipes is due to hoop-ring stresses resulting in longitudinal breaks.

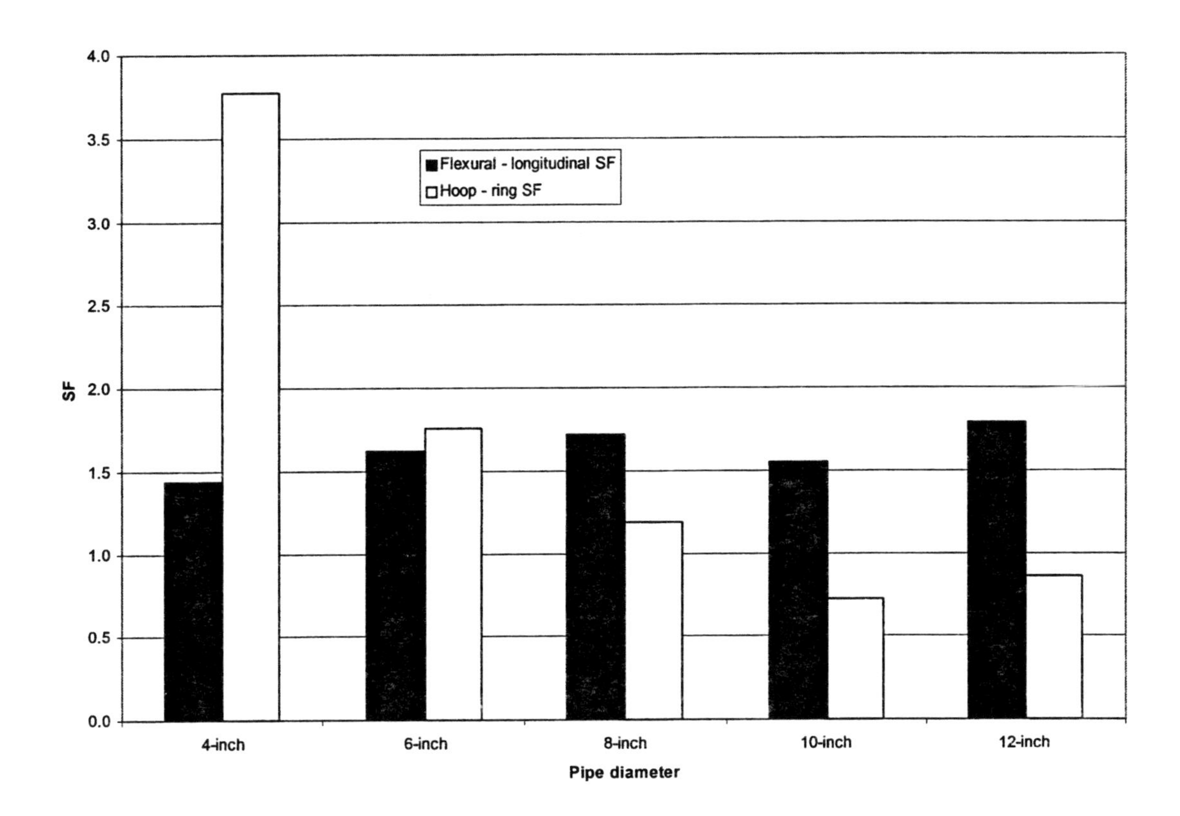

Figure 5.2 SF for RMOC test case for different pipe diameters

The effect of pipe age on the SF was also examined. The model assumes that the strength of a pipe decreases over time as a result of loss of wall thickness due to external corrosion. Therefore, the SF should decrease over time since the loads are assumed to remain the same. Figures 5.3 and 5.4 show the SF_{f-l} and SF_{h-r}, respectively, plotted against installation year of the pipe.

As expected, SF are generally lowest for the oldest pipes in the inventory. In Figure 5.3 the SF_{f-l} clearly increases as the installation year approaches the present day. The distinct bands in Figure 5.3 reflect the grouping of SF sharing the same soil category. The SF_{h-r} shown in Figure 5.4 exhibit much more scatter than shown by the SF_{f-l} in Figure 5.4. The SF_{h-r} do show patterns similar to the SF_{f-l} in that they are generally lower for the oldest pipe, and are grouped by soil category.

Figure 5.3, and to a lesser extent Figure 5.4, also illustrates an interesting point related to pipe manufacturing standards. The $SF_{f\text{-}l}$ for pipes installed between 1939 and 1951 are considerably lower than expected. A closer look at the design standards for pipes explains why this occurs. Table 5.4 summarizes pipe wall thicknesses for 6- and 8-inch cast iron pipes manufactured until 1966. Wall thicknesses were generally and gradually being reduced as manufacturing techniques improved. However, they were dramatically reduced between 1939 and 1951, amounting to about a one-third reduction in wall thickness. It is unknown if this was due to anticipated manufacturing improvements or related to material shortages during the war. After 1951 the wall thicknesses were increased, although not quite to pre-1939 values.

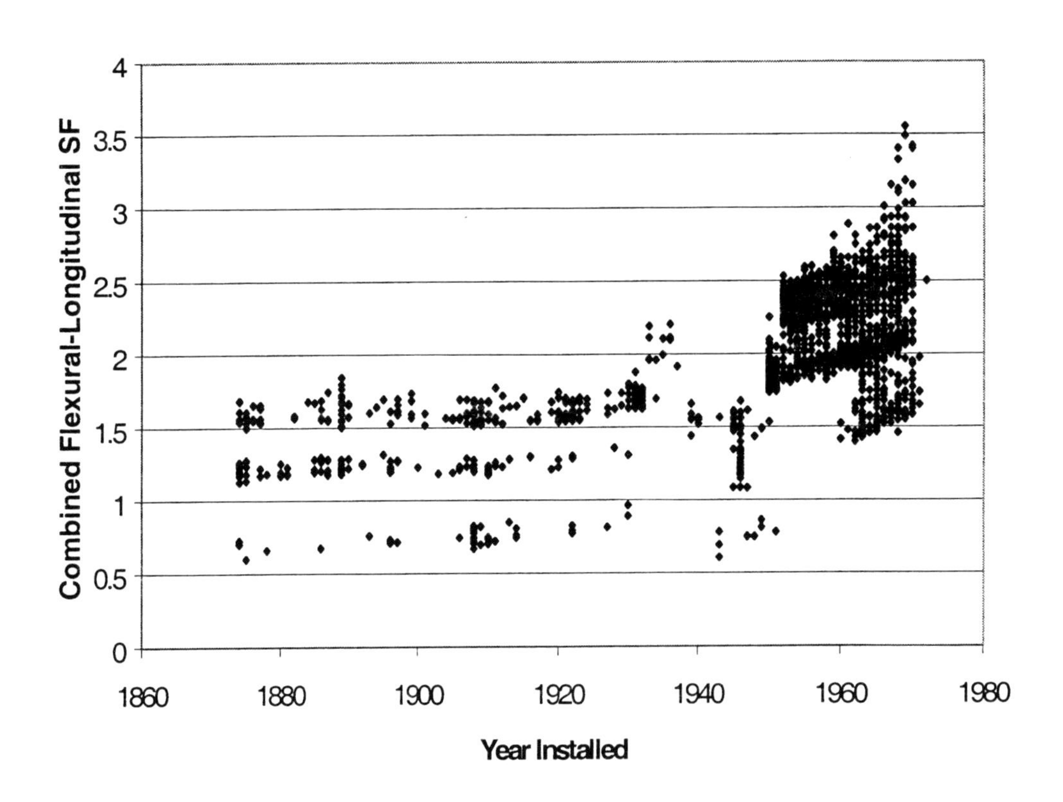

Figure 5.3 $SF_{f\text{-}l}$ for RMOC test case by installation year

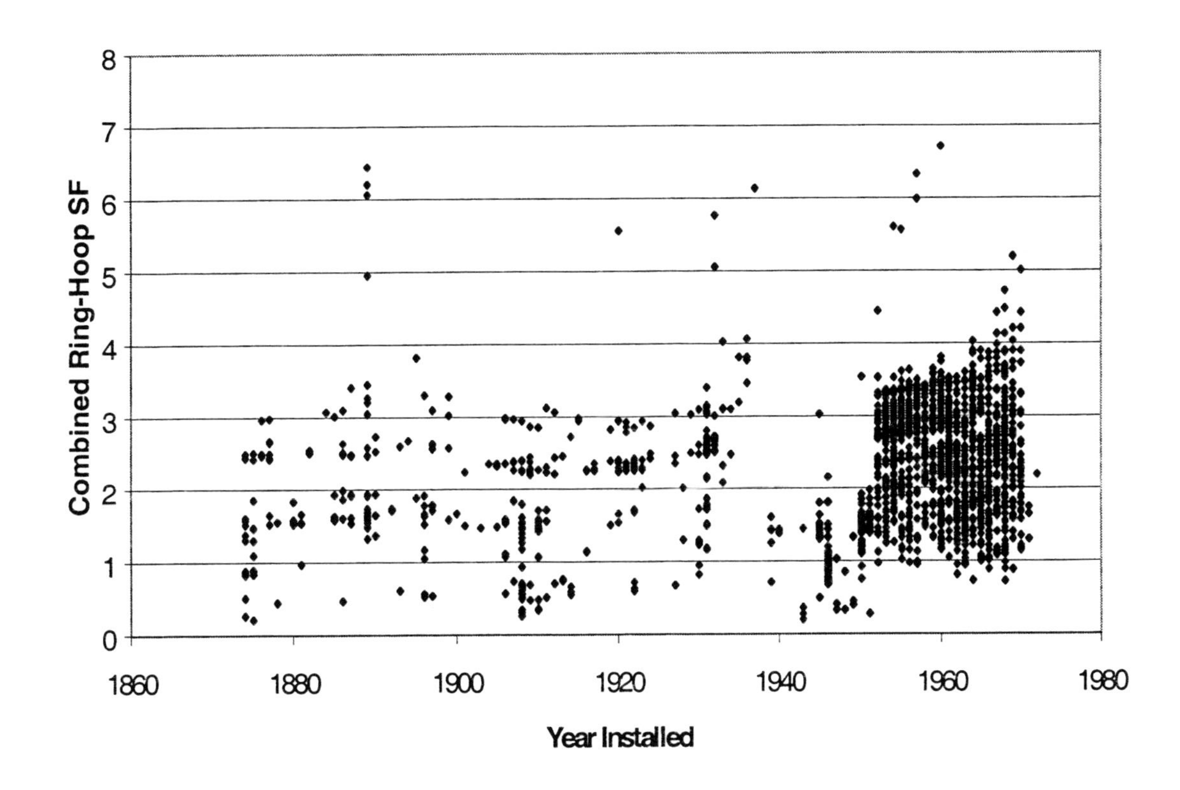

Figure 5.4 $SF_{h\text{-}r}$ for RMOC test case by installation year

Table 5.4

Cast iron pipe wall thicknesses over time

Pipe diameter, in.	Wall thickness, in.				
	Pre 1901	1901-1908	1909-1938	1939-1951	1952-1966
6	0.47	0.44	0.43	0.28	0.38
8	0.47	0.44	0.46	0.32	0.41

When the predicted safety factors are used to prioritize water mains for rehabilitation or replacement, the minimum value of the combined flexural-longitudinal safety factor, $SF_{f\text{-}l}$, and the combined hoop-ring safety factor, $SF_{h\text{-}r}$, should be used since this represents the likely failure mode for the individual pipes. Figure 5.5 shows the minimum combined safety factor ($SF_{f\text{-}l}$ and $SF_{h\text{-}r}$) versus pipe installation year for all pipes in the RMOC pipe inventory database.

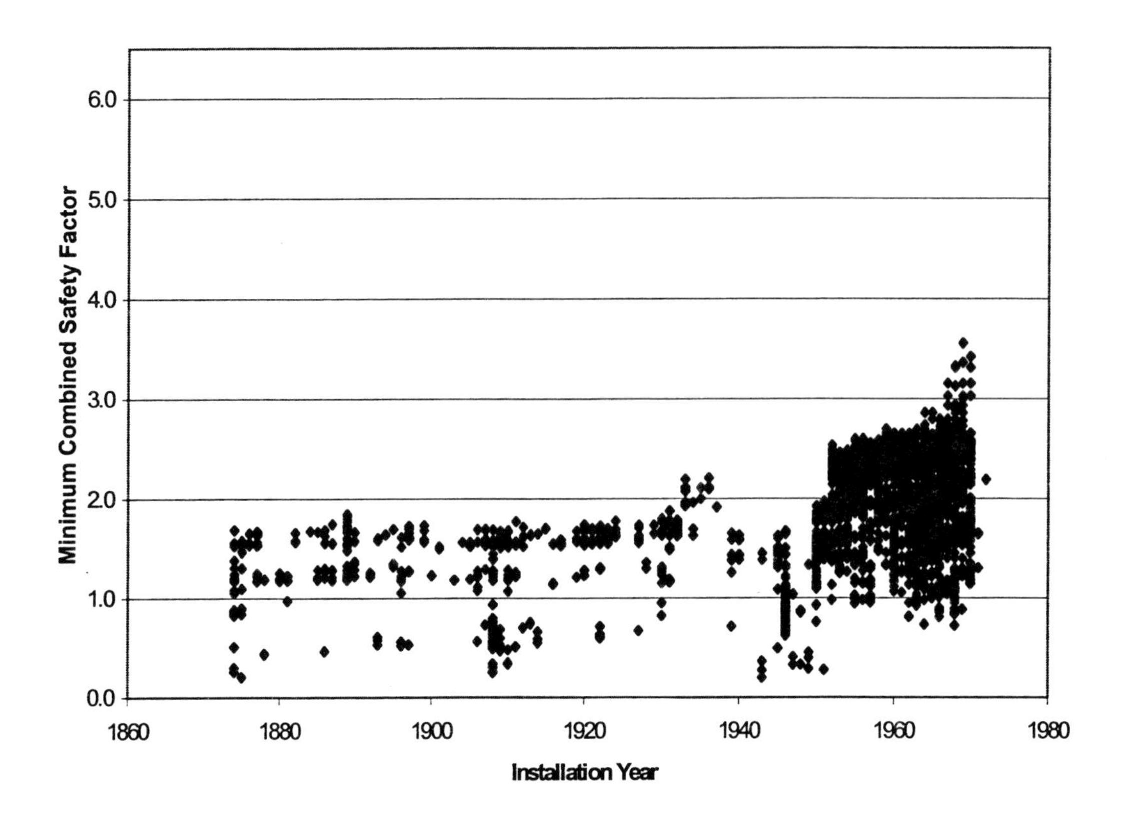

Figure 5.5 Minimum combined SF (SF_{f-l} or SF_{h-r}) for RMOC test case by installation year

Another way to examine the effect of pipe age on the SF is shown in Figure 5.6. This figure provides the average SF_{f-l} and SF_{h-r} for different groupings of pipe based on their age at failure.

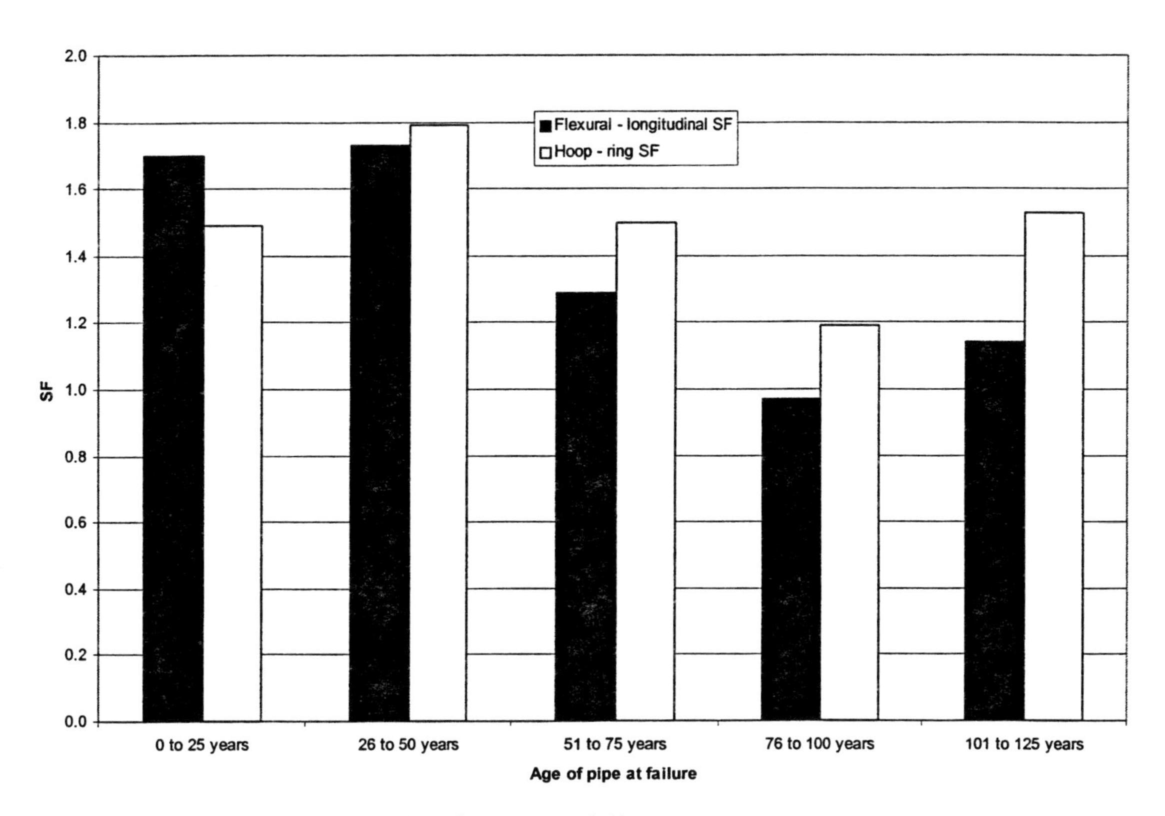

Figure 5.6 SF for RMOC test case by pipe age at failure

CONCLUSIONS

- The PBM was able to predict changes in the SF of cast iron pipe based upon age, soil type, and year of manufacture.
- Soil type had the most pronounced effect on the predicted SF for the pipes.
- There was greater variability in the predicted SF for combined ring and hoop stress than for combined flexural and longitudinal stress.
- The predicted flexural and longitudinal SF (SF_{f-l}) were slightly lower than the combined ring and hoop stress SF (SF_{h-r}) on average. However, the RMOC break history indicates that circumferential breaks (that would be caused by flexural and longitudinal stresses) occur much more frequently than longitudinal breaks (caused by ring and hoop stresses). Therefore, it would be expected that SF_{f-l} would be lower than SF_{h-r}.
- The PLM may be over-predicting the external loads for critical conditions used to calculate SF.

CHAPTER 6
FINDINGS AND RECOMMENDATIONS

Water main breaks are a continuing maintenance issue for all water utilities. Without effective water main renewal programs, it is likely that the rate of water main breaks will increase for many utilities into the future. Utilities must take advantage of the opportunities presented by water main breaks to gather the data necessary to assess the condition of pipes in their distribution systems. The key findings and recommendations of this project are provided below.

FINDINGS

This study evaluated the main break data collection practices of water utilities, and identified ways that these utilities could incorporate these data in developing water main renewal programs. A mechanistic model was developed that can help utilities prioritize cast iron water mains for renewal based on a comparison of anticipated loadings on the pipe versus estimated residual strength of the pipe. The key findings of this study are presented below.

- Estimates of distribution system renewal needs in the U.S. are between $77 and $325 billion over the next 25 years. Water utilities need to develop effective and practical water main renewal programs to meet this need.
- The AWWA WATER:\STATS database and a survey of North American water utilities conducted for this study indicate that the rate of water main breaks is between 20 and 25 breaks/100 mi/yr in North America. In Europe, the rate of water main breaks is approximately 80 breaks/100 mi/yr.
- Water main breaks provide an excellent opportunity for water utilities to gather information on the condition of pipes in their system. While it is apparent that water main breaks are a key component in planning a water main renewal program, only 70% of North American and 85% of European utilities indicated that they had computerized records of main failures.

- Useful applications of water main break data can be implemented even with only the most basic available data.
- A literature search identified a variety of methods used to prioritize pipes for replacement in a water main renewal program. Some methods used a scoring system to assign points to pipes based on various characteristics of the pipe and its environment. Others used economics to compare the costs of repair versus replacement. Still others attempted to predict the probability of future failures.
- Water main break related data can be classified as field, office, testing, and cost data. A comprehensive main break data collection program encompassing all four categories of data was recommended.
- A field test of the recommended data collection program identified several key factors that must be considered if a data collection program is to be successful. These factors included:
  - Field crew input is vital to the success of the effort.
  - Data collection forms must be customized for each utility.
  - The data collection form should be kept as simple as possible.
  - Proper handling and labeling of soil and pipe samples must be considered before the samples are collected.
  - The coordination of data collection must be assigned to a specific individual or group.
- Applications of water main break data range from simple trend analyses to sophisticated infrastructure planning models.
- A mechanistic model was developed to simulate the deterioration of cast iron pipe in the ground. Loads applied to the pipe were also calculated based upon the environment of the pipe. Knowing the remaining strength of the pipe and loads on the pipe, the model can calculate a SF for the pipe. Utilities can use the model to prioritize water main renewal programs by focusing on those pipes with the smallest SF.
- The mechanistic model could be a useful tool for prioritizing water main renewal programs, however it was apparent that calibration and field verification of model predictions were necessary to improve the usefulness of the model. Due to the

complex nature of corrosion and the fact that the mechanistic model was limited to external corrosion from corrosive soils, the mechanistic model is not a water main break prediction model and is not appropriate to be used in that way.

- Remaining wall thickness is the key parameter needed to estimate the residual strength of a pipe. The calculation of residual strength of a pipe based on remaining wall thickness is well established and reliable.
- Although it is believed that expansive soils can exert a significant load on buried pipes, there are no well-established methodologies for calculating these loads.
- The fracture toughness of a material is believed to be a key indicator of strength. Laboratory testing of fracture toughness for cast iron is not commonly performed. However, fracture toughness can be calculated from the more common laboratory measurements of tensile strength, ring stress, and modulus of rupture (flexural modulus) using Equation C.30 found in Appendix C.
- Standard main break repair practices (i.e., clamps) of most water utilities make it unlikely that pipe samples would be collected at every main break. Also, the costs of laboratory testing and other costs related to structural testing of pipes suggest that only selected pipes be thoroughly tested.
- Structural testing of pipe samples taken from water main breaks can provide useful data for calibrating mechanistic models. In particular, measurement of remaining wall thickness provides an important assessment of corrosion over time.

RECOMMENDATIONS

The following recommendations are made as a result of this study:

- All utilities should collect and maintain water main break data. The suggested data collection guidelines presented in Chapter 3 form a good foundation for a utility to use in developing their own data collection program.
- Water utilities should consider the purpose of their main break data collection program when deciding what types of data to collect. Even the simplest main break

data collection program can provide valuable input when planning distribution system renewal programs.

- Data collection programs must be coordinated with field crews in order to be effective. Every effort must be made to include field crews in the design of the data collection program. Without the cooperation of the field crews, the collection of data may be inconsistent, at best.
- A computerized database should be developed to manage and maintain the data collected during water main breaks. Analyses of main break data is greatly facilitated by having it available electronically.
- Water utilities should consider researching older paper records of main breaks, and incorporating them into a comprehensive water main break database.
- The main break data should be used in developing prioritized water main renewal programs. However, it must be recognized that main breaks alone do not necessarily determine the order in which pipes should be renewed.
- The mechanistic model developed in this study could be used by water utilities to prioritize cast iron water mains for renewal.
- Structural testing of pipe should be limited to selected pipes as circumstances and budget allow. However, measurement of wall thickness should be performed whenever possible.
- Water utilities should consider programs to assess the condition of their distribution system at times other than during main breaks. This is particularly true for soils data, since the "normal" characteristics of the soil can be changed significantly when the soil is wetted from water discharging from a broken pipe.
- Additional research is needed on the contribution of internal corrosion to the loss of wall thickness. A collaborative research effort to collect pipe coupons and test the specimens for thickness, tensile strength, modulus of rupture, and ring stress would allow better predictions by the SCM. Associated with this effort is the need to collect corresponding soil samples that should be tested for moisture content, bulk density, pH, and resistivity. This research would benefit all water utilities using cast iron pipe.

APPENDIX A
UTILITY RESPONSE TO LARGE WATER MAIN BREAK

Water utilities typically follow these specific steps in responding to a water main break:

1) Detection
2) Preliminary Assessment
3) Shutdown
4) Full Assessment
5) Repair
6) Test
7) Restore System Pressure
8) Return to Service
9) Restore Site In-Kind
10) Report

Detection, preliminary assessment and shutdown will occur almost simultaneously. The full assessment and determination of repair and test methods will be started well before the shutdown is completed. These steps are common for leaks as well as pipe breaks. Depending on the size of the break or leak and the number of customers affected, the number of subtasks and the extent to which these tasks are carried out and by whom will vary.

For example, a break on a large diameter transmission main may be easily and quickly detected and located, but the assessment of the condition of other portions of the large main and forensics as to what caused the break will require time. Most likely with a large main break, outside experts will be called in to assist with the assessment and method of repair. On the other hand, a small main break on a main that is part of the distribution system may take more time to detect and locate, but the cause and method of repair may be obvious and carried out by the utility's repair crew the same day.

DETECTION

Detecting a main break or leak can occur in the following ways:

- the public or local law enforcement may report water in the street
- instrumentation/ SCADA readings of pressure, flow, or tank level may indicate to an operator that something is wrong
- both may happen at almost the same time

Most frequently, it is the public that makes the report and it is most important that the utility actively promote and inform the public where to call if they notice what they believe to be a water main break. It is also important that the utility responds properly to a call by providing 24 hour monitoring, call tracking, updates, and repair crew response. The detection, monitoring, and response should also be coordinated with the local law enforcement agency.

PRELIMINARY ASSESSMENT

When a break is detected, it must first be located. Large breaks are usually located easily, unless there is an unusual circumstance such as a break at a water crossing. For small breaks, leak detection procedures and equipment may be required.

The utility must establish an on-site presence and communication with authorities, and a repair contractor and consultant if needed. For small breaks, this may be a repair crew and a door-to-door notification. For a large break, this requires a nearby operations center and public notification using radio and television of known service impacts. The public notification should include the following critical information:

- Is there loss of service?
- Are there water restrictions
- Is there a concern for contamination?

Preliminary assessment includes identifying the involvement of other utilities for water to supplement service needs, or electric, gas, telephone, and cable that may have been affected by the break. Simultaneously, with on-site presence, the utility must establish traffic control with local law enforcement.

The utility should assess the magnitude of the break and impacts to the system. Simultaneously, the size, type, age, depth, and type of restraint of the affected main must be determined. The map location and any record drawings, shop drawings, and installation schedules should be retrieved and reviewed.

During the preliminary assessment, the following forces will be mobilized: utility forces for shutdown and repair, an emergency contractor, if needed, and consultants as needed for assistance with securing the work site, repair design, and materials/ forensic specialists.

SHUTDOWN

The efforts to shutdown the affected main will occur simultaneously with the preliminary assessment. For small breaks, this may only involve locating and closing the isolating valves. For large breaks, the need to reduce pumping to reduce flow and velocities must be considered before valves are operated closed. Reduced pumping to accomplish the shutdown may run contrary to operator attempts to maintain system pressure and service.

It may be necessary to fall back to valves farther away from a large break so that flow near the break is reduced before the nearest isolating valves are closed. This will widen the portion of the service area affected by the shutdown. Other considerations include the location of the isolation valves with respect to the break and excavation to ensure that the main and valve have sufficient thrust restraint, and whether the valves that must be operated to secure the shutdown are operable or if the shutdown must be broadened.

FULL ASSESSMENT

The utility must use the opportunity following the shutdown to improve all the measures instituted to this point. This includes improved traffic control once the repair area is secured and contained, repair equipment is on-site, and some public roads or intersections can be returned to at least part-time use. Communications must be improved and the utility must update the public about how the repair is proceeding. It is important that the information be limited to only the facts and that there is one spokesperson. There should be no conjecture and as information becomes known, the public should be updated.

It may be necessary to install a trench box or sheeting for safe work access, depending on the depth of the excavation. As the work area is secured, the utility and its consultants must examine the pipe and determine whether the repair will be an in-place repair, will require a repair fitting, or will require new pieces of pipe. The pipe joints, deflection, and method of thrust restraint must be considered. At the same time, the method to restore the pipe bedding under the portions of the existing pipe to remain and under the repair areas must be determined with the method and materials for backfill, whether the sheeting will remain in place, and the detail for temporary and final pavement repair.

Once the method of repair is determined, the materials for the repair must be secured and/ or fabricated. It may be necessary to change the initial assessment of the repair method and/ or materials depending on what items are immediately available. This is crucial for the repair to be completed successfully and expeditiously. It is this step that tests the prior experience and resourcefulness of the individuals responding for the utility, its contractor, and consultant.

With the repair method finalized and the repair materials being secured and fabricated, the utility must update the authorities and the public as to the progress that is being made and the anticipated schedule for the repair, testing, and restoration of service if it can be known at this time.

Also, this time may be used for inspections to determine the condition of other portions of the pipe by visual inspection or remote camera observed by experts trained to identify pipe distress from certain conditions on the inside of the pipe. The need to use experts cannot be overstated first and foremost for safety during the confined space entry and second because benign conditions can look like pipe distress to the untrained eye and actual conditions of pipe distress can be overlooked.

Forensic inspection is important for large main breaks to help determine the cause and experts have a keen sense of what to observe. Large main breaks create large site disturbances and evidence of what was the initial cause of the break can be difficult to find even for the individual experienced in the construction of water mains. For main breaks of significance it is important to engage the services on-site of an expert in the particular pipe material(s), fabrication, testing, and construction and that has the experience of looking for and analyzing evidence on several past main breaks for the most probable cause.

REPAIR

As the repair items are brought to the site, the repair can be completed. It is essential that continuous, on-site inspection be provided to be able to make immediate field decisions. This ensures that the repair methods so carefully identified as part of the full assessment are carried out as intended. Continuous, on-site presence also ensures that all pieces that must be saved for further analysis are secured and stored and not disposed of.

As the repair proceeds, it is again important that updated information be provided to the public and the authorities as to progress and the anticipated schedule for the repair and restoration of service.

TEST

Testing may include weld inspection and a pressure test. Weld inspection by a materials specialist may be visual for both size and quality, or may use dye penetrant or magnetic particle techniques to enhance the visual inspection.

Flushing and disinfection precedes the pressure test. The pressure test is usually limited to visual inspection of the repair area while under system pressure since the isolation valves are usually not tight and leakage testing with pressures above system pressure cannot be conclusive. It is important that consideration be given to thrust restraint during the pressure test if visual observation is to be used to make sure the affected sections of the pipe are properly restrained during the test.

RESTORE SYSTEM PRESSURE

This should be done in a controlled manner and for a large main break often requires planning to determine the sequence to restore service so that system pressures are maintained or changed slowly as service is restored. Again, it is important that the public and authorities be updated as to the progress towards restoring service and for the public to know what to expect as service is restored. For example, the public should be informed about the need to boil water, if necessary.

RETURN TO SERVICE

The isolating valves will be opened in a planned sequence. A boil water notice, if originally instituted, may remain in effect until bacteriological test results are obtained and are negative. The system will then be brought back to normal service and the public and authorities must be updated again as to the status.

RESTORE SITE IN-KIND

This is an important part of the repair that often affects how the utility's response to the crisis is perceived. Every effort should be made to restore the site to its original or better condition.

REPORT

Assessment of the break cause and repair response brings necessary closure to the event. For a small break, this may include materials testing, assessment of the repair response time, number of customers affected, and a decision as to whether the main should continue in service, be rehabilitated or replaced. Other issues to be discussed include the effectiveness of the shutdown and valve operation to decide if valves need to be repaired or replaced or if new valves should be installed to improve the flexibility for shutdown.

For large breaks, materials testing, formal written reports on probable cause and a critique of the repair response must be prepared with recommendations for new guidelines and response procedures. The future actions of the utility will be guided by these determinations. Clearly, a large break on a critical transmission main may be tolerated once as an accident, but a second break on the same line in close proximity or in a short time on another line, will receive criticism whether the utility is effectively managing its infrastructure to serve its customers. For large mains, an assessment must be made to determine:

- the remaining service life of the pipe
- if the main can continue to serve as it is
- if the emergency repair should be considered temporary
- additional action that must be taken to make the repair permanent
- if the main should be rehabilitated or replaced

APPENDIX B

EXAMPLES OF WATER UTILITY MAIN BREAK DATA COLLECTION FORMS

Actual main break data collection forms used by water utilities are provided on the following pages. Each utility should consider its own data collection needs, as well as soliciting input from field crews responsible for collecting the data, in developing their own forms.

DENVER WATER
Corrosion Engineering
Leak Inspection

Location	Date	Size

Type of Pipe		Cause of leak	
☐ Asbestos Cement	☐ Galvanized Iron	☐ Beam Break	☐ Service Insulator
☐ Coated Steel	☐ Grey Cast Iron	☐ Corrosion of Pipe	☐ Tap Pullout
☐ Concrete	☐ Plastic	☐ Coupling	☐ Valve
☐ Ductile Iron	☐ Other ______	☐ Lead Joint	☐ Other ______
	______	☐ Leadite Joint	______
		☐ Mechanical Joint Bolt Corrosion	

Resistivity	How Resistivity Determined
General Soil Condition	Location of Resistivity

Pipe Information

Grey Cast, Ductile & Galvanized Iron	Coated Steel Pipe
☐ Polyethylene Wrapped ☐ During Installation ☐ At Time of Leak	Type of Protective Coating ______ Condition of Coating ☐ Good ☐ Fair ☐ Poor
☐ Pitting ______ : ______ % of wall thickness	☐ Disbonding ☐ Bacterial Deterioration ☐ Moisture present under coating ☐ Bare spots ☐ Pockets ☐ Sagging
☐ Graphitization ______ : ______ % of wall thickness	Pitting ______ : ______ % wall thickness Pipe-to-soil potential ______
☐ Pipe Condition ☐ Good ☐ Fair ☐ Poor	Location ______

Location of leak relative to:

Insulating couplings ______

Foreign utilities ______

Steel in concrete ______

Copper pipe ______

Other ______

Possible stray current (from appearance of metal) ______

Remarks: ______

______ Overtime Hrs. ______ Foreman ______ Inspected By

5010/7/80

Figure B.1 Example leak inspection sheet

COMMISSIONERS OF PUBLIC WORKS OF THE CITY OF CHARLESTON S.C. — CPW

CPW WORK ORDER Work Order # **R**

☐ MAIN-BREAK ☐ REPAIR ☐ REPLACE Wk Rqst # ________

WATER DISTRIBUTION DEPARTMENT

Addr/Loc:	Day: _____ PPE: __/__/__
Cross St: ft. N-S-E-W of PL/CL	Date:___/___/___
	Start time: ____: ____am/pm
Area: ______________ Subdivision: ______________	Finish time: ____: ____am/pm

Work Description: ____

☐ Main-Break ☐ Hydrant ☐ Valve ☐ Blow-off ☐ Other ☐ Patch: ____ Water ____ Sewer

Failure Date: __/__/__ Hydrant/Valve No: _-_-_ Size/vo ____	Tuberculation ID: ____" Polywrapped: Y / N Sample: Y / N
Make/Model: ________/________ Year: _____ Depth: ____'____"	Corrosion: ☐ Outside ☐ Inside - ☐good ☐ fair ☐ poor

Cause of Failure:			Type of Failure:			Patch information:	
☐ Traffic	☐ High pressure	☐ Temp. chng	☐ Blow-Out	☐ Sleeve	☐ Joint - Lead	Pvmt. Type:	Size Cut:
☐ Corrosion	☐ Water Hammer	☐ Deterioration	☐ Split-Bell	☐ Clamp	☐ Joint - Gasket		
☐ Misuse	☐ Leak w/ Operation		☐ Longitudal	☐ Pinhole	☐ Split @ Corp		
☐ Defective	☐ Improper bedding		☐ Circumferential (transverse)				
☐ Contractor	☐ Other: ________		☐ Other: ________				

Failure Location: ft. N-S-E-W of P/L-C/L | Main Type: (DI,CI,PVC,AC,LJ) - O.D. " | Patch No:

Soil Found: ☐ Sand ☐ Clay ☐ Silt ☐ Other: ________ | Backfill: ☐ Original ☐ Other - Describe: ________

MATERIALS INSTALLED:					**MATERIALS INSTALLED:**				
Item:	Size:	Qty.	Make/model	Cost	Item:	Size:	Qty.	Make/model	Cost
					MATERIALS FAILED / REMOVED:				

Crew No:	Hours:		On	Sequence No:	Equip. Used:			
Crew Members:	Reg.	OT	Call	Resp. Code: 011	Type:	size:	Hrs:	Cost:
				Job No:	Crew Truck			$
				Acct No:	Dump Truck			$
				Reg. No:	Backhoe			$
				PO No:	Compressor			$
					Pump			$
Pay code totals: 11-	31-	32-		Other-				$
Cost totals: 120-	130 -			140 -				$

Other Utilities Damaged: Y / N | Were Utilities Located? Y / N | Located Properly? Y / N | Locate No:

Type(s) Of Utilities Damaged :

Comments:

Hydrant Flushing: Y/N			Chlorination: Y/N
Time:	Pitot Read:		Reading:_______ ppm
(Min.)	PSI	GPM	Turbidity: Y/N
			Reading: _______ ntu
			Anode Inst: Y/N
Total gallons:			Resist: ______ ohm / in

As-built Drawing:

North Arrow:

PMP-Task Cd: ______ Units: ______ Pnts: ______

3-C Cards Issued: Y / N - No. Issued: ________

Damage Claim Info: ☐ Bill ☐ No Bill

Name of Company:

Address: City: St:

Driver/Forman: Zip:

License No: State:

Vehicle No: Phone: ( ) -

General Contractor:

Locate Requested? Y/N - Req. No: ________

Was Utility Located? Y / N - Date located: / /

Signature of Responsible Party: X ________________ | Approved By (Locator's Signature): ________

Repair complete: Y / N | Damage Comments:

Additional Comments:

Supervisor's Signature: | District Supervisor's Sign:

Figure B.2 Example work order sheet

PG&W PENNSYLVANIA GAS AND WATER COMPANY
511-517 EAST NORTHAMPTON STREET
WILKES-BARRE, PA 18711

YEAR ☐ SURVEY TYPE ☐ PAGE NUMBER ☐

DATE: ☐-☐-☐ MONTH DAY YEAR

LEAK CLASSIFICATION: A ☐ B ☐ C ☐

COMPANY LEAK ☐ CUSTOMER LEAK ☐

WATER LEAKAGE CONTROL REPORT

ESTIMATE LOSS ☐ (GALLONS/MINUTE)

DIVISION ☐ OPERATING AREA ☐ (SEE REVERSE SIDE FOR VALID AREAS)

CITY ☐

ADDRESS ☐ SOURCE OF SUPPLY ☐

SOURCE OF REPORT
- SURVEY ☐
- COMPLAINT ☐

SURVEY TYPE
- MAIN LINE ☐
- HYDRANT ☐
- SERVICE LINE ☐
- COMPLAINT ☐
- PREPAVE ☐
- OTHER ☐

LOCATION OF PIPE
- STREET ☐
- CURB LAWN ☐
- YARD ☐
- R-O-W ☐
- OTHER ☐

INDICATION OF LEAK
- SONIC ☐
- SURFACE WATER ☐
- LOW PRESSURE ☐
- OTHER ☐

LEAK OCCURS ON
- MAIN ☐
- SERVICE ☐

LOCATION OF LEAK
- HYDRANT ☐
- VALVE ☐
- PIPE ☐
- TAP ☐
- BLOW OFF ☐

PIPE COVER
- CONCRETE ☐
- ASPHALT ☐
- BRICK ☐
- SOIL ☐

PINPOINTED BY ☐ DATE ☐ ☐ ☐

REPAIRED BY ☐ DATE ☐ ☐ ☐

REMARKS: ____________________

SIGNATU[RE]

SD-10-007 (6/91) ORIGINAL N855

Figure B.3 Example water leakage control sheet, page 1

WATER LEAK REPAIR DATA

ORDER NO. ☐☐☐☐

TYPE OF LEAK	TYPE OF REPAIR
☐ - (MB) MAIN BREAK (CIRCUMFERENTIAL)	☐ (PC) PIPE CLAMP
☐ - (BO) BLOW - OUT	☐ (SP) SPLIT SLEEVE
☐ - (SM) SPLIT MAIN	☐ (BJ) BELL JOINT CLAMP
☐ - (BJ) BELL JOINT	☐ (MC) MAIN REPLACEMENT (CAPITAL)
☐ - (HY) HYDRANT	☐ (MM) MAIN REPLACEMENT (MAINT.)
☐ - (MV) MAIN VALVE	☐☐☐☐☐ FLEET
☐ - (SP) SERVICE PIPE	☐ (MO) MAIN OTHER ☐☐☐☐☐☐☐☐☐☐☐☐☐☐
☐ - (CS) CURB STOP	☐ (SR) SERVICE REPLACEMENT
☐ - (OT) OTHER ☐☐☐☐☐☐☐☐☐☐☐	☐ (PR) PARTIAL REPLACEMENT (SERVICE)
☐☐☐☐☐☐☐☐☐☐☐	☐ (RE) SERVICE RETIREMENT
	☐ (SO) SERVICE OTHER ☐☐☐☐☐☐☐☐☐☐☐☐☐

MATERIALS USED: ☐☐☐☐☐☐☐☐☐☐☐☐☐☐☐☐☐☐☐☐☐☐☐☐☐☐☐☐☐☐☐☐☐☐

NUMBER OF LEAKS REPAIRED: ☐☐

LEAK REPORT PAGE NUMBER KEY

YEAR REPAIRED	TYPE SURVEY	YEAR REPORTED	PAGE NUMBER	LEAK CLASS	LOCATION	CAUSE
☐☐	☐☐	☐☐	☐☐☐☐	☐	☐	☐☐

TYPE SURVEY

ML – MAIN LINE
SL – SERVICE LINE
HY – HYDRANT
PR – PREPAVE
OC – OUTSIDE CONST.
SP – SPECIAL
CM – COMPLAINT

LEAK CLASS

A = "A" LEAK
B = "B" LEAK
C = "C" LEAK

LOCATION

M = MAIN
S = SERVICE

CAUSE

CR = CORROSION
TP = THIRD PARTY
OF = OUTSIDE FORCE
CD = CONST. DEFECT
MD = MATERIAL DEFECT
OT = OTHER

DIVISION

51 = SCRANTON WATER RATE AREA
61 = SPRINGBROOK WATER RATE AREA

OPERATING AREAS

NORTH

CARBONDALE
OLYPHANT
SCRANTON

CENTRAL

NANTICOKE
PITTSTON
WILKES-BARRE

Figure B.4 Example water leakage control sheet, page 2

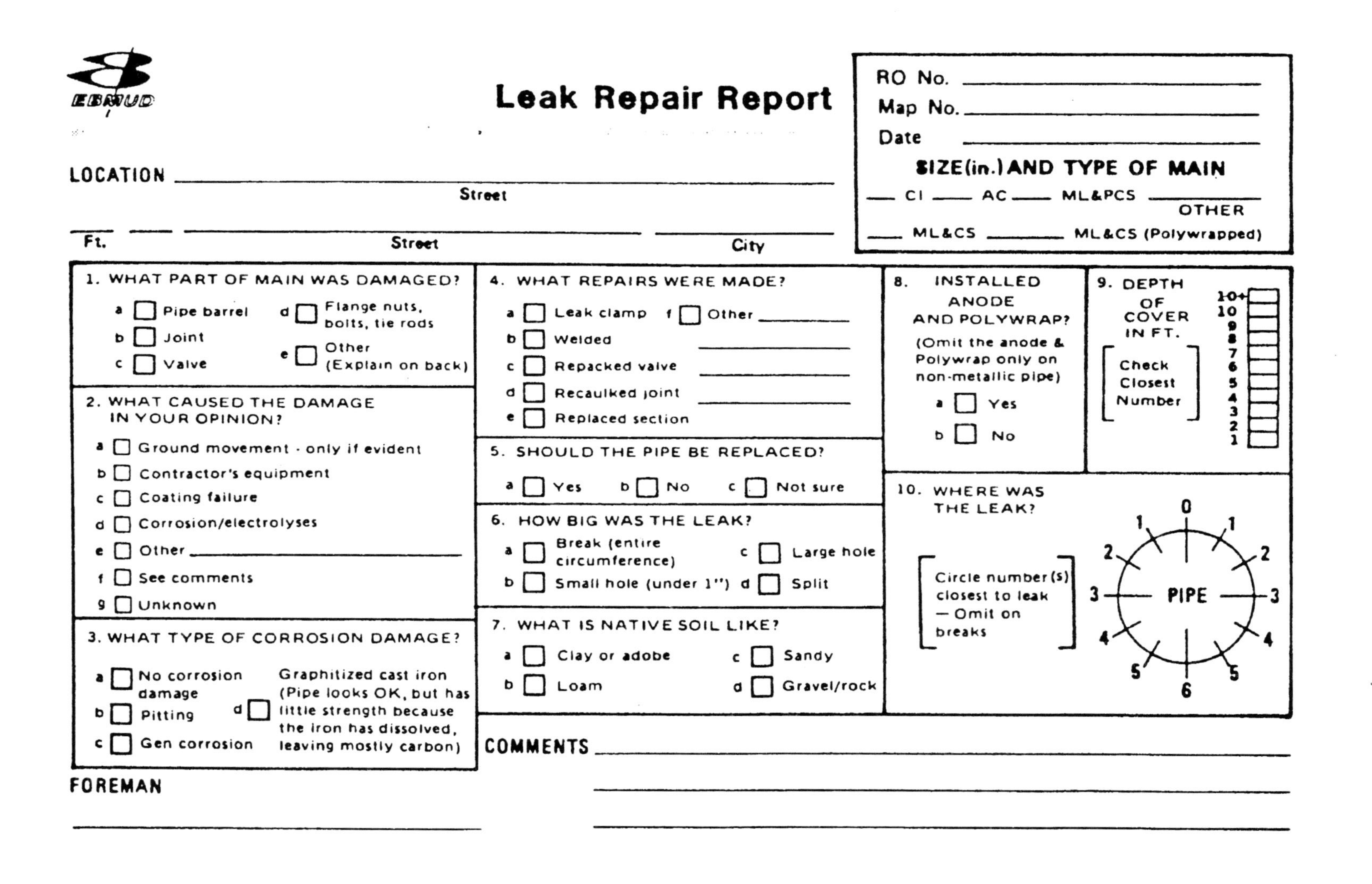

EBMUD

Leak Repair Report

RO No. ______
Map No. ______
Date ______

SIZE(in.) AND TYPE OF MAIN

___ CI ___ AC ___ ML&PCS ______ OTHER
___ ML&CS ______ ML&CS (Polywrapped)

LOCATION ______ Street

___ ___ ______ Street ______ City
Ft.

1. WHAT PART OF MAIN WAS DAMAGED?
   - a ☐ Pipe barrel
   - b ☐ Joint
   - c ☐ Valve
   - d ☐ Flange nuts, bolts, tie rods
   - e ☐ Other (Explain on back)

2. WHAT CAUSED THE DAMAGE IN YOUR OPINION?
   - a ☐ Ground movement - only if evident
   - b ☐ Contractor's equipment
   - c ☐ Coating failure
   - d ☐ Corrosion/electrolyses
   - e ☐ Other ______
   - f ☐ See comments
   - g ☐ Unknown

3. WHAT TYPE OF CORROSION DAMAGE?
   - a ☐ No corrosion damage
   - b ☐ Pitting
   - c ☐ Gen corrosion
   - d ☐ Graphitized cast iron (Pipe looks OK, but has little strength because the iron has dissolved, leaving mostly carbon)

4. WHAT REPAIRS WERE MADE?
   - a ☐ Leak clamp
   - b ☐ Welded
   - c ☐ Repacked valve
   - d ☐ Recaulked joint
   - e ☐ Replaced section
   - f ☐ Other ______

5. SHOULD THE PIPE BE REPLACED?
   - a ☐ Yes b ☐ No c ☐ Not sure

6. HOW BIG WAS THE LEAK?
   - a ☐ Break (entire circumference)
   - b ☐ Small hole (under 1")
   - c ☐ Large hole
   - d ☐ Split

7. WHAT IS NATIVE SOIL LIKE?
   - a ☐ Clay or adobe
   - b ☐ Loam
   - c ☐ Sandy
   - d ☐ Gravel/rock

8. INSTALLED ANODE AND POLYWRAP? (Omit the anode & Polywrap only on non-metallic pipe)
   - a ☐ Yes
   - b ☐ No

9. DEPTH OF COVER IN FT. [Check Closest Number]
   10+ ☐ 10 ☐ 9 ☐ 8 ☐ 7 ☐ 6 ☐ 5 ☐ 4 ☐ 3 ☐ 2 ☐ 1 ☐

10. WHERE WAS THE LEAK? [Circle number(s) closest to leak — Omit on breaks]

COMMENTS ______

FOREMAN ______

Figure B.5 Example leak repair report sheet

EPCOR
essential elements for living

FAILURE REPAIR FORM

Work Request # ______ Form Initiated By: ______ Date: ______

Facility Type (1): Bend	Facility Type (2): ______	Facility Type (3): ______
Facility # (1): ______	Facility # (2): ______	Facility # (3): ______

CONCERNS INVESTIGATION

Nearest Intersection St: ______ Av: ______ Concern #: ______ Cadastral #: ______

Facility Align. on: ______ at: ______ Hydrant Out of Service: ___ Investig. By: ______

People Out of Water: ___ No. of Houses: ______ No. of Apartment/Suites ______ No. of Businesses: ______ Hose Hookup: ___

Water Tank: ___ Water Serv. - Applic. #: ______ Water Serv. Addr: ______ City/Private: ______

Street Surface: ______ Pavement Cut No.: ______ Date/Time Reported: ______

Date/Time Confirmed: ______ Main Break W.O. #. ______ Date/Time Off: ______

Investigative Comment: ______

FAILURE DETAIL

Apparent Cause of Failure: (1) ______ (2) ______ (3) ______

Failure Description: (1) ______ (2) ______ (3) ______

Cond. of Facility: (1) ______ (2) ______ (3) ______

Remarks: ______

REPAIR DETAIL --- Dig Up: N

Work Order: ______ Task: ______

Rep. Start: ______ Rep. Compl: ______ Damage Prop: ___ Damage Utility: ___ Frost Depth: ______

Crew No. (1): ______ Foreman: ______ Crew No. (2): ______ Foreman: ______

Actual Location On: ______ At: ______ Watermain/Hydrant Lead Alignment: ______

Clamp/Cplg Location (1): ______ Clamp/Cplg Location (2): ______ Watermain Depth: ______

Facility I.D.	Repaired or Replaced Facilities	Manufacturer	Model	Size
		Repaired / Replaced Parts		

Flushing: ___ No. of Minutes: ______ Disinfection: ___

Field Water Quality Test: ___ Turbidity: ______ Chlorine: ______

Lab Water Quality Test: ___ Physical Chemical Sample No: ______ Microbiological Sample No: ______ Re-Sample: ___

Number of Anodes Installed: ___ Anode Type: ______ Weight: ______ Testing Station: ___ *(If Yes, Detailed Installation Report Attached)*

Recommision Date: ______

Remarks: ______

Reviewed By: ______ Data Entered By: ______

Date: ______

Figure B.6 Example failure repair form sheet

Mar-24-00 11:33am From-P S W C ENGINEERING DEPT. 610-645-1055 T-039 P.02 F-203

PHILADELPHIA SUBURBAN WATER COMPANY

MAIN LEAK REPORT

DIVISION ______ MUNICIPAL # ______ DATE ______

MAIN SIZE ______ (Inches) * MAIN TYPE ______

SHUT DOWN (Y/N) ______ ADDRESS ______

LOCATION FROM: ______ LOCATION TO: ______

TIME OFF ______ TIME ON ______

CUSTOMERS ______ NOTIFIED (Y/N) ______

PLATE #1 ______ PLATE #2 ______

EXTENSION # ______ * MAIN LEAK TYPE ______

* OTHER LEAK TYPE ______ * REPAIR TYPE ______ FT. ______

DEPTH ______ FOREMAN ______

OF VALVES CLOSED ______ REPAIR TIME (HRS/MIN) ______

APPARENT PIPE CONDITION – GOOD ( ) FAIR ( ) POOR ( )

SHUTDOWN – GOOD ( ) FAIR ( ) POOR ( ) DAMAGE CAUSED – SLIGHT ( ) MODER. ( ) SEVERE ( )

* CAPITAL YES ( ) NO ( )

DIRECTION & DISTANCE TO VALVE OR HYD. ______

COMMENTS ______

* ENTER CODE FROM MASTER CODE SHEET *

FORM L-115 REV. 8/93

Figure B.7 Example main leak report sheet

MAR-23-00 THU 10:56 UW M AND S ENGINEERING FAX NO. 2016344220 P.02

United Water New Jersey
Distribution System Repair Report

Entry date ______

I. Engineering office data

Plan no. ____ LCA no. ____ System pressure ____

Main installed by
☐ Our const. ☐ Outside contr.

Repair information

CSB no. ____ Date ____

II. Field data

Location

Municipality ______ Leak location ______
Street no. ______ Dist. in ft. from curb ______
Nearest interset. st. in feet ______

Pipe/service information

☐ Water main — Pipe size (in.) ____ Ext. no. ______
☐ Service pipe — Service size (in.) ____ Service no. ______

Problem/repair information

Material:
☐ C.I.–unlined
☐ C.I.–cem. lined
☐ D.I.–cem. lined
☐ Prestressed conc.
☐ Asbestos cem.
☐ Steel–unlined
☐ Steel–cem. lined
☐ Steel–coal tar lined
☐ Brass
☐ Lead lined galv. iron
☐ Copper
☐ Lead
☐ Tubeloy
☐ Galvanized iron
☐ Plastic
☐ Unknown

Type of problem:
☐ Joint leak–lead
☐ Joint leak–leadite
☐ Joint leak–MJ
☐ Joint leak–POJ
☐ Joint leak–PCCP
☐ Joint leak–PCECP
☐ Damaged rubber gasket
☐ Broken bell
☐ Damaged or broken offset
☐ Circumferential break
☐ Longitudinal break
☐ Hole
☐ Break–unknown

Location of joint leak 1 2 3 4 5 6 7 8

Materials used and work done:
☐ Repair clamp
☐ Bell joint clamp
☐ Recalk lead joint
☐ Tighten mechanical joints
☐ Pipe and dresser couplings
☐ None
☐ Other–explain below
☐ Service renewal
☐ Service repair
☐ Pipe only

Should the pipe be replaced?
☐ Yes
☐ No
☐ Not sure

Repair time required ____ **hours**

Conditions/installation

Damaged by:
☐ Outside contractor
☐ Our contractor
☐ Other utility
☐ Unknown

Pavement type:
☐ Reinf. conc.
☐ Belg. blk.
☐ Bit. conc.
☐ Asphalt/conc. base

Other Utility:
☐ O&R–elect.
☐ O&R–gas
☐ N.Y. tel.
☐ AT&T
☐ N.J. Bell

Traffic cond.:
☐ High
☐ Medium
☐ Low

Cause of damage:
☐ Pipe/rock contact
☐ Corrosion
☐ Settlement
☐ Other–explain below
☐ Unknown

Digging cond.:
☐ Dry
☐ Undergr. water
☐ Ongoing leak
☐ Incompl. shutdown
☐ Other

Corrosion:
☐ Internal
☐ External

Type of soil:
☐ Rock
☐ Sand/gravel
☐ Clay
☐ Fill
☐ Loam
☐ Organic
☐ Other

Depth of coverage in feet
☐ 2'
☐ 3'
☐ 4'
☐ 5'
☐ 6'
☐

Remarks ______

______

______

______ Foreperson

Figure B.8 Example distribution system repair report sheet

DC Water and Sewer Authority
Bureau of Water Services

Site Report for Pipeline Repair / Inspection

Job No. ____________ Permit No. ____________

1. Work Assignment Complaint: ____________

Reported by ☐ Employee ☐ Customer ☐ Other Name: ____________ Date______ Time ______ Assigned to: ____________ Date______ Time ______	Crew Chief: ____________ Arrival time: ________ Departure: ________ Water shutoff: ________ Restored: ________	Address: ____________ Counter map: ____________ Description of exact location: ________ ____________ ____________ ____________

2. Repair / Inspection Report (see back side of this sheet)

3. Valve Operations

Number of valves closed to shut main down: ____ (Identification: ____________

____

Number of valves opened: _____ Identify valves in incorrect position or in need of repair: __

____

4. Pipe Information Diameter: _____ Depth of cover: _____ (check more than one if necessary)

Pipe material ☐ (CI) Cast iron ☐ (DI) Ductile iron ☐ (St) Steel ☐ (Cn) Concrete ☐ Other ________	Joint ☐ Lead ☐ Rubber push-on ☐ Mechanical ☐ Flanged ☐ Restrained Type ________ ________ ☐ Not seen ☐ Other ________	External protection ☐ Bitumen ☐ Plastic wrap ☐ Cathodic protect ☐ None visible ☐ Other ________ Internal protection ☐ Cement lined ☐ None visible Tuberculation depth ______inches	Surface conditions ☐ Heavy traffic ☐ Light traffic ☐ Outside of street Ground type ☐ Soil ☐ Clay ☐ Fill ☐ Rock Bedding? ☐Yes ☐No

5. Failure information ☐ Pipe failure ☐ Valve leak ☐ Hydrant

(check more than one if necessary) ☐ Non-repair inspection ☐ Other ________

Pipe failure type: ☐ Circular break ☐ Longitudinal break ☐ Hole blowout (large) ☐ Small hole (less than ½") ☐ Pipe joint ☐ Joint (previous repair) ☐ Tapping point	Located on: ☐ Pipe ☐ Fitting Leakage rate: ☐ Slight ☐ Moderate ☐ Severe	Comments of possible failure cause: ________ ____________ ____________ ____________ ____________ ____________

6. Samples (tag with Job no., date, address, pipe material and size) ☐ Pipe coupon / ring (circle) ☐ Soil

7. Closure repair documented on BWS counter maps: signed by ____________ date ______

Figure B.9 Example site report for pipeline repair / inspection sheet, page 1

DC Water and Sewer Authority
Bureau of Water Services

Site Report for Pipeline Repair / Inspection

Job No. ___________ Permit No.___________

8. Field Reports (Crew chief reports from each inspection or repair crew.)

Date/Name	Report

Sketch area

Figure B.10 Example site report for pipeline repair / inspection sheet, page 2

DC Water and Sewer Authority
Bureau of Water Services

Office Pipe Sample Inspection Report

Job No. _______ Permit No.______ Office inspector: __________

Date: ________________

9. Pipe Material Confirmation Inside diameter:____ inches Wall thickness: ____ inches

Pipe material
☐ (CI) Cast iron ☐ (DI) Ductile iron ☐ (St) Steel ☐ (Cn) Concrete ☐ Other ______

10. Inspection of Pipe Interior

Tuberculation: ☐ Whole surface ☐ Most of surface ☐ None of surface ☐ Soft (easy to remove) ☐ Hard (difficult to remove)	Intact lining ☐ Whole surface ☐ Most of surface ☐ None of surface	Other Observations: __________
Max height of tuberculation: _______ inches		
Remaining clear bore: _______ inches		

11. Inspection of Pipe Exterior (wire brush and scape if necessary) Measured ext diam: ____ in.

External protection intact over:	Pitting:	Other Observations: ________
☐ Whole surface	☐ None visible	
☐ Most of surface	☐ Some ☐ A lot	
☐ None of surface	Max pit depth _____ mm.	
Encrustation on:	Graphitisation:	
☐ Whole surface	☐ None visible	
☐ Most of surface	☐ Some ☐ A lot	
☐ None of surface		

12. Inspection of Cut Edges Wall thickness: largest ____ in. smallest: ______ in.

Graphitisation:	Graphitisation is:	Pitting:
☐ None	☐ Hard	☐ None
☐ Interior (max depth __ mm.)	☐ Soft	☐ Interior (max depth __ mm.)
☐ Exterior (max dep ___ mm.)	(easily removed)	☐ Exterior (max dep ___ mm.)
☐ Whole thickness		☐ Whole thickness

13. Laboratory Testing Requested? ☐ Yes (Lab: ______________________) ☐ No

Describe or list testing requested:________________________________

Figure B.11 Example office pipe sample inspection report sheet

BHC
An Aquarion Company

FIELD DATA FOR MAIN BREAK EVALUATION

DATE OF BREAK: ____________ TIME: ____________ A.M. ____________ P.M.

TYPE OF MAIN: ____________ SIZE: ________ JOINT: ________ COVER: ____FT ____ IN

THICKNESS AT POINT OF FAILURE: ____________ IN.

NATURE OF BREAK: ☐ Circumferential ☐ Longitudinal ☐ Circumferential & Longitudinal ☐ Blowout ☐ Joint ☐ Split at Corporation ☐ Sleeve ☐ Miscellaneous: ____________

APPARENT CAUSE OF BREAK: ☐ Water Hammer (surge) ☐ Defective Pipe ☐ Corrosion ☐ Deterioration ☐ Improper Bedding ☐ Excessive Operating Pressure ☐ Differential Settlement ☐ Temperature Change ☐ Contractor ☐ Miscellaneous: ____________

SIDE OF STREET: ____________ ☐N ☐S ☐E ☐W

TYPE OF SOIL: ____________

ELECTROLYSIS INDICATED: ☐ Yes ☐ No CORROSION: ☐ Outside ☐Inside

DEPTH FROST: ____________IN. DEPTH OF SNOW: ____________IN.

MAIN BREAK EVALUATION

WEATHER CONDITIONS / TEMPERATURE: ____________

WATER TEMPERATURE *(by office)*: Temp. ________°F

TYPE OF MAIN: ____________ CLASS OR THICKNESS: ________ LAYING LENGTH: ______ FT.

DATE LAID: ____________ OPERATING PRESSURE: ____________

BACKFILL: ☐ Native Material DESCRIBE: ____________

☐ Bank Run Sand & Gravel ☐ Gravel ☐ Sand ☐ Crushed Rock ☐ Other: ____________

TOWN: ____________

STREET: ____________

CROSS STREET NO.1: ____________

CROSS STREET NO.2: ____________

MSLINK #: ____________ *(to be completed in office)*

LENGTH OF SHUTDOWN (HRS.): ____________

NUMBER OF CUSTOMERS WITHOUT SERVICE: ____________

REPORTED BY: ____________

DAMAGE TO PAVING AND/OR PRIVATE PROPERTY: ____________

REPAIR MADE (Materials, Labor, Equipment): ____________

REPAIR DIFFICULTIES (if any): ____________

REPAIR CONTRACTOR: ____________

Figure B.12 Example field data for main break evaluation sheet

___03/27/00 MON 15:53 FAX 203 337 5839 BHC ENG DEPT 003

PIPELINE STATUS REPORT

C NO. ________
Date ________
Permit No. ________
SIZE(In.)AND TYPE OF MAIN
__CI __AC __PVC __OTHER
__DI

LOCATION ______________________________
Street

___ ___ ______________________ ______________________
ft Street City

1. WHAT PART OF MAIN WAS DAMAGED?
a ☐ Pipe barrel
b ☐ Joint
c ☐ Valve
d ☐ Flange nuts, bolts, tie rods
e ☐ Other (Explain on back)

2. WHAT CAUSED THE DAMAGE IN YOUR OPINION?
a ☐ Ground movement - only if evident
b ☐ Contractor's equipment
COMMENTS ______________________

3. WHAT TYPE OF CORROSION DAMAGE?
a ☐ No corrosion damage
b ☐ Pitting
c ☐ Gen corrosion
d ☐ Graphitized cast iron (pipe looks OK, but has little strength because the iron has dissolved, leaving mostly carbon)
Comments ______________________

4. WHAT REPAIRS WERE MADE?
a ☐ Leak clamp
b ☐ Joint Clamp
c ☐ Replaced section
COMMENTS ______________________

5. SHOULD THE PIPE BE REPLACED?
a ☐ Yes b ☐ No

6. WHAT IS NATIVE SOIL LIKE?
a ☐ Clay or adobe
b ☐ Loam
c ☐ Sandy
d ☐ Gravel/rock

7. HOW BIG WAS THE LEAK?
a ☐ Break (entire circumference)
b ☐ Small hole (under 1")
c ☐ Large hole
d ☐ Split

Indicate size and location of leak

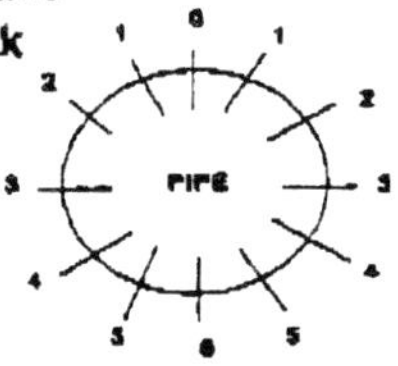

PIPE (clock positions 1–6)

8. DEPTH OF COVER IN FT.
X ______________

9. HOW MUCH BUILD UP WAS THERE?
Fill in circle to represent build-up (each circle represents 1 inch)

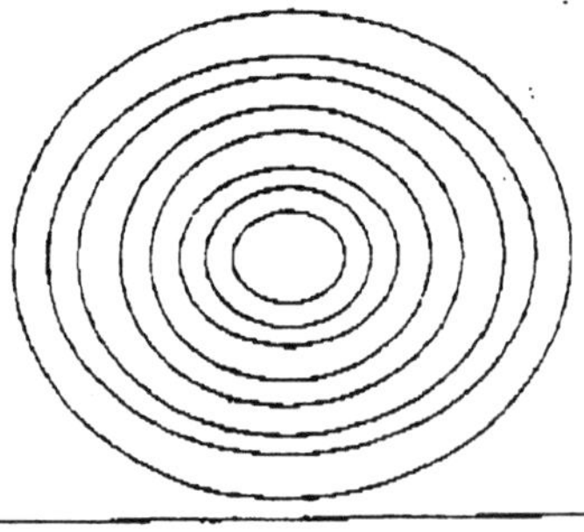

10. UNDERGROUND OBSTRUCTIONS (check and define)
a ☐ Water ______________
b ☐ Gas ______________
c ☐ Telephone ______________
d ☐ Electric ______________
e ☐ Sanitary Sewer ______________
f ☐ Storm Sewer ______________
g ☐ Other ______________

11. MATERIALS USED
____

12. TYPE OF PAVEMENT REMOVED — opening size
a ☐ Asphalt ______________
b ☐ Oil-Gravel ______________
c ☐ Concrete ______________
d ☐ Concrete-Asphalt ______________
e ☐ Concrete Curbing ______________
f ☐ Asphalt Curbing ______________

COMMENTS ______________________

FOREMAN ______________________

Figure B.13 Example pipeline status report sheet

APPENDIX C

DETAILS OF MECHANISTIC MODELS

INTRODUCTION

Chapter 4 described a variety of applications using data collected during water main breaks. These ranged from a simple accounting of the number of main breaks to sophisticated modeling of the vulnerability of pipe to failure for the purpose of prioritizing pipes for renewal. A mechanistic model of this type was developed in conjunction with this project. The details of the model are described in this appendix, while the actual use of the model is described in a User Guide in Appendix D.

DEVELOPMENT OF MODELS

The following sections describe the calculation procedures for the Pipe Load Module (PLM), Pipe Deterioration Module (PDM), Statistical Correlation Model (SCM), and Pipe Break Model (PBM).

Pipe Load Module (PLM)

The PLM module accounts for all the loads acting on a buried pipe. These loads fall into two main categories: external (traffic or live load, earth load, and frost load, expansive soil load, temperature induced expansion/contraction load) and internal (working pressure and surge pressure). Though the expansion and contraction load has been classified as external load, it may be induced by the change in temperature of the water flowing through the pipe. The loads related to pipe failure and their interaction is shown in Figure C.1.

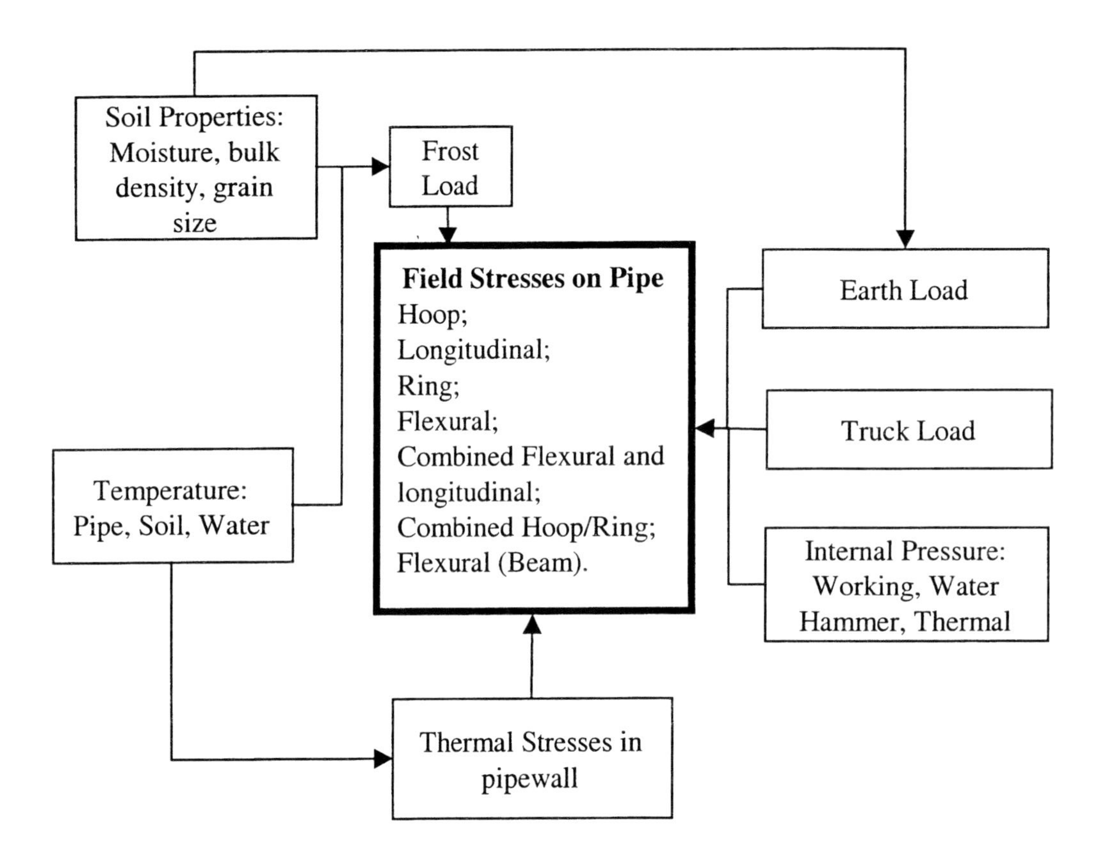

Figure C.1 Pipe Load Module

Earth Load (W_e)

The computation of earth loads on a pipe is dependent on the type of installation or ditch condition used. For ditch condition, the pipe laid is relatively narrow trench backfilled to original ground surface (CIPRA 1978). Trench width at the top of the pipe determines the earth load. The trench may be widened above the top of the pipe for ease in installation without increasing the load on pipe. For the model, the bedding type was chosen as type A. Under bedding type A, the pipe is laid on flat bottom trench, backfill not tamped.

The calculations are based on the recommended procedure outlined by ANSI/AWWA C101 as described in the Cast Iron Pipe Research Association Handbook (CIPRA). The general earth load per lineal foot is given by:

$$W_e = C_d \gamma B_d^2 \quad \text{(C.1)}$$

where W_e = earth load, lb/linear ft

C_d = a calculation coefficient

γ = unit weight of soils overlying pipe, lb/ft^3

B_d = width of trench at top of pipe, ft

The actual values of C_d and B_d depend on the depth of cover, earth pressure, soil friction, and the type of ditch condition used.

$$C_d = \frac{1 - e^{-(2K_r\mu' \frac{H}{B_d})}}{2K_r\mu'} \quad \text{(C.2)}$$

where

K_r = ratio of active horizontal pressure at any point in the fill to the vertical pressure which causes the active horizontal pressure.

μ' = coefficient of sliding friction between fill materials and sides of trench
$K_r\mu'$ = 0.130 for standard calculations.

H = height of fill above the top of pipe, feet.

Traffic Load (W_t)

The ANSI/AWWA C101 also describes the calculation of expected loads due to trucks depending on the following conditions: pavement type (unpaved, flexible, or rigid pavement); one or two passing trucks; wheel load; and impact factor. For unpaved or flexible pavement, the equation used is:

$$W_t = CRPF \quad \text{(C.3)}$$

and for rigid pavement,

$$W_t = KDPF \tag{C.4}$$

where W_t = truck superload (lb/lin ft)

C= surface load factor for unpaved of flexible pavement. C values can be obtained from the CIPRA Handbook (Tables 1-11 and 1-12)

R =reduction factor, accounting for the part of the pipe directly below the wheels (See Table 1-13 in CIPRA Handbook)

P= wheel load (lb) (Standard, P = 9,000 lb)

F= Impact factor (Standard, F = 1.50)

K = surface load factor for rigid pavement (see Table 1-14 in CIPRA Handbook)

D = outside diameter of pipe (ft) (see Table 1-15 in CIPRA Handbook)

Expansive Soil Load

Expansive soil loads can generate significant stresses on buried pipes. Although there are no standard or widely accepted methodologies for estimating the magnitude of expansive soil loads on pipes, one methodology is presented here as a reference. Because of the great uncertainty in estimating expansive soil loads, it has not been incorporated into the PLM. The following is taken from CIPRA(1978). An expansive soil is any soil that swells upon wetting and shrinks upon drying. The expansive soils are generally clay soils with size ranging from less than 1 micron to 2 microns. The following field observations are suggested in identifying an expansive soil: (1) become hard on drying and fissures extensively, (2) becomes very sticky when wetted, (3) plastic over a wide range of water content, (4) have a soapy or slick feeling when rubbed between fingers, (5) fine grained with little sand or coarse material, and (6) simply known as clay. Expansive soil is considered to cause beam breaks. Among the expanding group of clay known as smectite group, the most common member is the Montmorillonite (Brady, 1990; Spangler and Handy, 1982). Issa (1997) derived the following set of equations through experimentation to help predict the swelling pressure based on the moisture content of a soil.

$$P_{sw} = \alpha_1 (w_L - 46) \text{ when } 9.4\% \leq w_0 \leq 16.2\%; \tag{C.5}$$

$$P_{sw} = \alpha_2 (w_L - 56) \text{ when } 21.4\% \leq w_0 \leq 27.5\%; \tag{C.6}$$

$$P_{sw} = \alpha_3 (w_L - 77) \text{ when } 32.5\% \leq w_0 \leq 33.1\%; \tag{C.7}$$

where $\alpha_1 = 0.0300$ MPa (4.347 psi)

$\alpha_2 = \alpha_3 = 0.0245$ MPa (3.549 psi).

W_L = liquid limit = water content above which the soil readily becomes a liquid.

W_0 = initial soil water content and water content is the ratio of weight of water to weight of solids expressed in percent

P_{sw} = the swelling pressure, Mpa

Because the range of the initial water content in Eq. (5) coincides with the shrinkage limit (water content at which the soil moves from the solid to semisolid state, that is, the volume begins to increase because of addition of water) for Montmorillonite (Lambe and Whitman 1976) and for a conservative estimate, Eq.(5) is used to calculate the swelling pressure. The liquid limit will have to be assessed from laboratory tests following the ASTM-D4318 test procedure. From Lambe and Whitman (1976) using the largest liquid limit value of 710% corresponding to Montmorillonite (with Na exchangeable ion), we obtain the greatest expansive soil pressure from eq. (5) as 2,886 psi; whereas, for Kalinite which does not expand has the greatest expansive soil stress of 56.5 psi based on eq.(5) for a liquid limit of 59%.

Frost Load (W_{fadj} or W_{frz})

The PLM accounts for frost load in one of three ways: (1) based on Monie and Clark (1974), DIPRA (1984), and Fielding and Cohen (1988) the frost load can be taken to be twice the earth load; or (2) an explicit user-supplied frost load magnitude; or (3) the frost load calculation from Rajani and Zhan (1996). The default option in the PLM is to use twice the earth load.

The frost load computed from Rajani and Zahn is adjusted as follows:

$$W_{fadj} = \frac{Fd_{max}}{Fd_c} * W_{frz} \tag{C.8}$$

where

W_{fadj} = the adjusted frost load, lb/ft

W_{frz} = the frost load computed from the Rajani and Zahn equation, lb/ft

Fd_{max} = the maximum frost depth supplied by the user, in

Fd_c= the frost depth computed from the Rajani and Zahn equation, in

According to Spangler and Handy (1982), an estimate of the freezing depth can be obtained from the Stefan equation as Fd = C_t Sqrt[k_fF/L] in which: Fd = depth of freezing in, m(ft); C_t = a constant =13.1 for SI units and 6.93 for English units; k_f = thermal conductivity coefficient of frozen soil, watts per meter-degree kelvin(Btu/sq.ft-hr-deg.F/ft) (0.8-2.1 for clay and 1.7-4.0 for sand); F = freeze index is the accumulated daily freeze index defined as the positive daily temperature difference below freezing; L = latent heat of fusion of water in soil kilo Joules/cubic meter = 340[percent water content][dry unit weight of soil in kilo Newtons/cubic meters]; replace 340 by 1.43 for lb/cubic ft. and corresponding L in Btu/cubic feet; note 1 Btu/sq.ft-hr-deg.F/ft = 1.73 Watts/meter-deg.kelvin).

Besides computing the frost load, the model also provides the user with an option to specify a frost load that may be used in the pipe analysis.

Internal Pressure (p)

This is the actual working pressure of the pipe defined by the ultimate demand the pipe is designed to meet. The longitudinal and hoop stresses resulting from working pressure are calculated below.

Water Hammer Pressure (P)

Water hammer pressure results from sudden stopping of flow in the pipe. Its effect depends on how quickly a valve is opened or closed. In the design of pipes, the CIPRA manual

Table C.1

Allowances for surge pressure for cast iron pipe design

Pipe size (in)	Surge pressure (psi)
3-10	120
12-14	110
16-18	100
20	90
24	85
30	80
36	75
42-48	70

Source: CIPRA 1978.

(1978) recommends a standard allowance for water hammer pressure based on the pipe size. These allowances are given in Table C.1 and are the default values for the model.

The surge pressure resulting from a rapid change in water velocity can be calculated from the following water hammer equations:

$$a = \frac{4660}{\sqrt{1 + \frac{kd}{Et}}} \tag{C.9}$$

and

$$P = \frac{aV}{2.31g} \tag{C.10}$$

where a = velocity of pressure wave, ft/sec

k = fluid bulk modulus, 300,000 psi for water

d = pipe diameter, in.

E = modulus of elasticity of the pipe, 1,500,000 psi for cast iron

t = pipe wall thickness, in.

P = surge pressure, psi

V = maximum velocity change, ft/sec (taken to be 2 ft/sec)

g = acceleration of gravity, 32.2 ft/sec^2

Thermal Expansion and Contraction ($\sigma_{l,T}$)

When exposed to a temperature differential, materials tend to expand or contract. During winter, the stress induced in the pipe wall due to thermal contraction of the pipe material emanates as a tensile stress because of the resistance offered by the surrounding soil. Similarly, during summer compressive stresses develop. In general the winter tensile stress is considered to be more severe because cast iron is weaker in tension than in compression (http://www.nrc.ca/irc/newsletter/v2no1/water.html). The thermal stress is calculated as:

$$\sigma_{1,T} = E\alpha\Delta T \tag{C.11}$$

where $\sigma_{l,T}$ = longitudinal stress due to thermal contraction, psi

E = modulus of elasticity of the pipe, 1,500,000 psi for cast iron

α = coefficient of thermal contraction, 6.26E-6 ft/ft/^{o}F

ΔT = temperature change, ^{o}F

Load-Induced Stresses on Pipes

External loads, internal loads, and temperature changes create various components of stress on the pipe wall, including ring stress, hoop stress, tensile stress, and flexural (bending) stress. These were illustrated in Figures 4.9 and 4.10. Methods used by the PDM for calculating the various stresses are described in the following sections.

Ring Stresses

Ring stresses are caused by external loads and temperature changes as described below.

Ring stresses due to external loads. This is the stress that is induced circumferentially in the pipe wall analyzing the pipe section as a circular beam. For any thickness and external load, this stress can be computed by (CIPRA 1978):

$$\sigma_\theta = \frac{0.0795w(d+t)}{t^2} \qquad \text{(C.12)}$$

where: σ_θ = ring stress in the pipe wall, psi

t = net thickness, in.

d = nominal inside pipe diameter, in.

w = external load on the pipe without internal pressure but includes, earth, truck, frost, and expansive soil loads (lb/linear foot)

Ring stresses due to water temperature change. According to Timoshenko and Goodier(1987), the ring stress in pipe due to temperature change, $\sigma_{t,\theta}$

$$\sigma_{t,\theta} = \frac{\alpha E \Delta T}{2(1-\nu)\log(b/a)}\left[1 - \log\frac{b}{r} - \frac{a^2}{(b^2-a^2)}\left(1+\frac{b^2}{r^2}\right)\log\frac{b}{a}\right] \qquad \text{(C.13)}$$

where: a = distance from center to inner wall, in

b = distance from center to outer wall, in

ν = Poisson ratio

r = distance of any point from the center, in

α= linear expansion coefficient of pipe, ft/ft

E = modulus of elasticity of the pipe, 1,500,000 psi for cast iron

ΔT = Temperature difference between the inner and outer walls taken to be the water temperature change in °F

Total Ring Stress

The total ring stress is the sum of the ring stress due to external loads and the ring stress due to thermal contraction.

$$\sigma_{\theta,total} = \sigma_{\theta} + \sigma_{t,\theta} \tag{C.14}$$

Hoop Stress

Hoop stress results from internal pressure in the water main. The pipe needs a certain minimum thickness to withstand the hoop stress when the pipe has no over burden load. This stress is computed for two cases: (1) working pressure and (2) water hammer pressure. These are added together to get total hoop stress. The hoop stress is computed as:

$$\sigma_h = \frac{Pd}{2t} \tag{C.15}$$

where σ_h = hoop stress, psi

t = net thickness, in.

d = nominal inside pipe diameter, in.

P = water pressure, psi

Total hoop stress is calculated from the water pressure. The individual components of hoop stress can be calculated separately by substituting working pressure or surge pressure

Combined hoop and ring stress. Crushing failure of pipe under external load is buckling caused by ring flexural stress in the pipe wall beneath the zone of load application. The critical flexural stress occurs at the inner surface of the pipe wall, which is where the critical hoop stress can develop under large internal pressures. To account for the pipe failure under the effect of both external loads and internal pressure, the following loading formula is provided by CIPRA (1978):

$$\frac{w^2}{W^2} = \frac{(P-p)}{P} \quad \text{(C.16)}$$

where p = internal pressure, psi

w = external load, lb/linear ft

P = internal pressure at bursting for the pipe, psi

W = ultimate crushing load, lb/linear ft

The variables p and w are any combination of internal pressure and external load respectively that will cause failure of the pipe. In using this equation, it is necessary to compute the W and P values first. The internal pressure at bursting, P is calculated from the bursting tensile strength, S (psi). The internal bursting pressure P (psi) is calculated by the PDM as:

$$P = \frac{2St}{d} \quad \text{(C.17)}$$

The crushing load, W can be calculated as:

$$W = \frac{\pi b R t^2}{3(d+t)} = \frac{Rt^2}{0.0795(d+t)} \quad \text{(C.18)}$$

in which for the second part PWD (1985b) took the width of the ring specimen, b = 12 inches and R = ring modulus of rupture (psi).

Let the internal pressure be p_i and let the total external load be w, then, the external load corresponding to the internal pressure p_i that will cause a failure is given by:

$$w = \sqrt{\frac{W^2(P-p)}{P}} \quad \text{(C.19)}$$

Longitudinal Beam Stress

Longitudinal beam stress represents a bending (flexural) stress on a beam simply supported at the ends with the central span of the beam unsupported. Deterioration of pipe bedding conditions can result in the uneven support of the pipe, causing it to be only supported at certain points along the pipe length. When such a situation arises, the pipe acts as a beam. The flexural stress or beam modulus of rupture in the pipe can be determined by:

$$\sigma_f = \frac{(15.28)w(D)L^2}{(D^4 - d^4)} \quad \text{(C.20)}$$

where σ_f = flexural stress at extreme fiber, psi

w = effective vertical load distributed along the unsupported pipe length, lb/linear foot

L = length of unsupported span, ft

D = outside diameter of the pipe, in.

d = inside diameter of the pipe, in.

The pipe thickness necessary to withstand the load under beam action is contained in the term D^4-d^4. The effective vertical load distributed along the unsupported pipe length, w is the sum of the traffic load, earth load, frost load and expansive soil load. Using the w value, the total flexural stress resulting from these loads can be computed.

Longitudinal pressure stress. In addition to internal pressure and ring loads, pipes are also subjected to longitudinal pressure stress resulting from hydrostatic conditions and changes in magnitude or direction of internal flow velocities. The longitudinal stress due to such pressures has a magnitude half as large as the hoop stress and can be defined as:

$$\sigma_l = \frac{pd}{4t} \tag{C.21}$$

where σ_l = the uniformly distributed longitudinal tensile stress due to internal pressure, psi

Longitudinal thermal stress due water temperature change. According to Timoshenko and Goodier (1987), longitudinal stress in pipe due to a rapid (e.g., less than 24–hour) temperature change, σ_t , is calculated as:

$$\sigma_t = \frac{\alpha E \Delta T}{2(1-\nu)\log(b/a)}\left[1 - 2\log\frac{b}{r} - \frac{2a^2}{(b^2-a^2)}\log\frac{b}{a}\right] \tag{C.22}$$

where a = distance from center to inner wall, in

b = distance from center to outer wall, in

ν = Poisson's ratio

r = distance of any point from the center, in

α= linear expansive coefficient of the pipe material, ft/ft/°F

E = modulus of elasticity of the pipe, 1,500,000 psi for cast iron

ΔT = Temperature difference between the inner and outer walls taken to be the water temperature change in °F

Longitudinal stress component of hoop stress. Hoop stress creates a component of longitudinal stress as a characteristic of the internal matrix of the pipe material. The ratio of the longitudinal component of hoop stress to hoop stress is the Poisson's ratio. The longitudinal component of hoop stress is calculated as:

$$\sigma_v = \nu\left(\sigma_{\theta,\text{total}} + \sigma_h\right) \tag{C.23}$$

where σ_v = longitudinal component of hoop stress, psi

$\sigma_{\theta,\text{total}}$ = total ring stress, psi

σ_h = total hoop stress, psi

ν = Poisson's ratio

Total longitudinal stress. In order to determine the combined effect of the longitudinal stress sources, the computed flexural stress is will be added to the longitudinal pressure stress and the longitudinal stress in pipe wall resulting from temperature change in water. The resulting value is the total expected longitudinal stress, σ_T.

$$\sigma_T = \sigma_f + \sigma_l + \sigma_t + \sigma_v \qquad \text{(C.24)}$$

Combined longitudinal and bending stress. Longitudinal and bending stresses act in the same direction at the extreme fiber at the bottom (outside) of the pipe and therefore may exceed the failure stress of the pipe at that location. The total longitudinal and bending stresses are simply added to evaluate this condition (refer to Figures 4.9 and 4.10).

Pipe Deterioration Module (PDM)

The ultimate goal of the PDM is to determine the reduction in the pipe's strength or load bearing capacity over time. The load capacity of the pipe is dependent on the wall thickness and the material strength. This module can be used to relate the factors that promote the deterioration of a pipe over time to the growth of pits in the pit wall. The module addresses only external corrosion due to corrosive soil conditions, as indicated by soil pH, resistivity, and soil aeration, although the "corrosion rate" parameter can be used to simulate loss of wall thickness from all sources, including internal corrosion. Data collected on pipe conditions over time will be used to develop a deterioration rate function for the pipes in the system. Figure C.2 provides a schematic description of various modes of pipe deterioration.

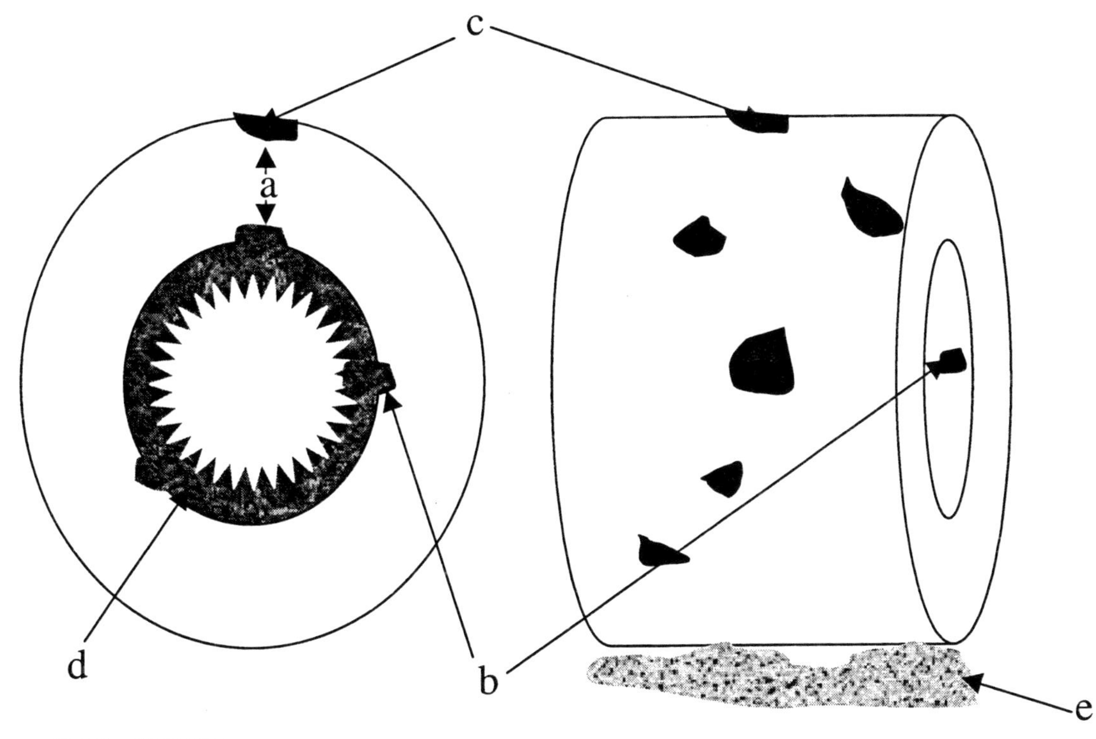

LEGEND

a = effective thickness

b = internal corrosion

c = external corrosion

d = tuberculation

e = disrupted bedding

Figure C.2 Illustration of reduced pipe wall thickness and loss of bedding support for a buried pipe

Corrosion pits, created by corrosion of the ferrous material, cause the tensile strength of the material to be reduced and have the effect of concentrating stresses in their vicinity. The structure of the module is described in Figure C.3.

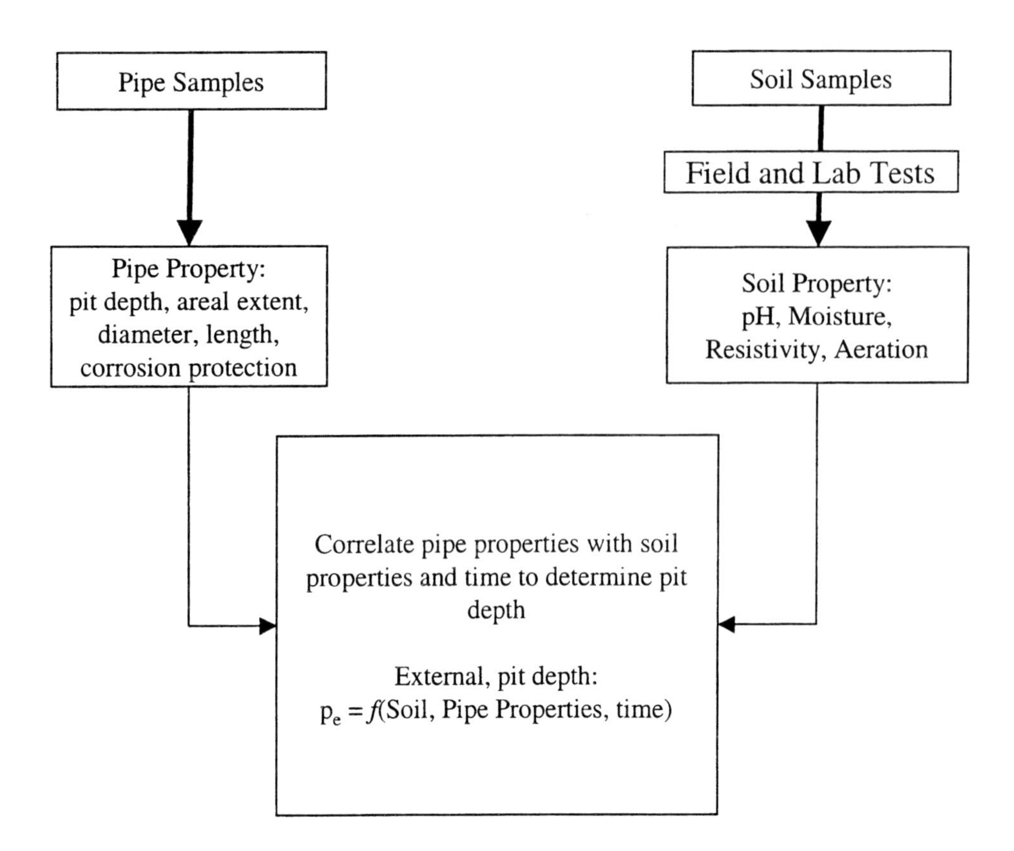

Figure C.3 Pipe Deterioration Model

External Corrosion

The PDM uses equations developed by Rossum (1969) to calculate the growth of external corrosion pits as a function of soil properties. The growth of pits due to corrosion on the outside walls of pipes is dependent on the soil property. For modeling the pit depth with time and soil environment and age, Rossum (1969) developed a set of equations. His equations are partly based on the extensive data collection effort by the National Bureau of Standards (NBS) (Romanoff 1957). The NBS buried specimens of different varieties in 47 soils starting in 1922. The specimens were removed from these sites in 1924, 1926, 1928, 1930, 1932, and 1934 and the last set of specimens from less corrosive sites in 1939. The results of exposures over the span of 2 to17 years, therefore, are available. Rossum took advantage of these results in developing his equations.

The Rossum pit growth equation is given by:

$$p = K_n K_a \left[\frac{(10 - pH)}{\omega} \right]^n T^n A^a \quad \text{(C.25a)}$$

where p = pit depth, mils

K_n and n = constants dependent on the soil aeration

pH = soil pH

ω = the soil resistivity, ohm/cm

A = the area of pipe exposed to the soil, ft^2

K_a and a = constants dependent on the pipe material

T = the time of exposure, years

The Rossum model is modified to better fit the data trend as follows:

$$p = 0.6183 * K_n K_a \left[\frac{(10 - pH}{\omega} \right]^n (T^n - 1) \quad \text{(C.25b)}$$

The model uses the modified Rossum's equation to compute the pit depth unless the user provides a corrosion rate. If the user provides a corrosion rate, the pit depth is computed by multiplying the age of the pipe and the corrosion rate. The values of n and K_n in the Rossum equation are determined as shown in Table C.2.

Table C.2

Soil aeration constants for use in Rossum corrosion pit growth equation

Soil Aeration	n	Kn
Good	0.16	170
Fair	0.33	222
Poor	0.50	355

Source: Rossum 1969. For cast iron K_a = 1.40, a=0.22

Statistical Correlation Module (SCM)

This module relates the residual strength of the pipe (lab sample) expressed in terms of the modulus of rupture, fracture toughness and the tensile strength with the corrosion status expressed in terms of the pit depth and environmental characteristics. With the aid of the combined PDM and SCM the relevant strength parameters of a pipe in terms of its environmental factors can be calculated. Figure C.4 shows a flow chart of fitting the SCM.

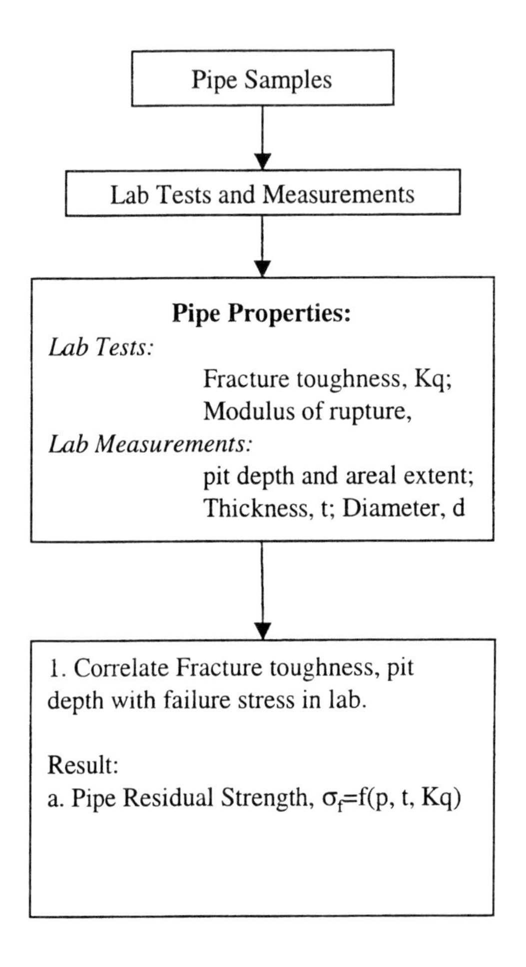

Figure C.4 Statistical Correlation Module

Fracture Toughness

The discussion related to fracture toughness closely follows Flinn and Trojan (1990). Failures occur in many components even when the operating stress is well below the yield stress.

Generally such failures are due to stresses in particular regions that have been amplified by the presence of holes, cracks, and other discontinuities. The capacity of a particular flaw to cause failure depends on the material property called fracture toughness. However, the stress concentration at the edge of the flaw depends on the geometry of the flaw and the geometry of the component but not on the properties of the material. An equation that relates the fracture stress and the fracture toughness is:

$$\sigma = \frac{K_{IC}}{Y\sqrt{\pi a}} \tag{C.26}$$

where K_{IC} = Fracture toughness , psi$\sqrt{\text{in}}$

σ = nominal stress at fracture, psi

a = a measure of crack length, in

Y = a dimensionless correction factor that accounts for the geometry of the component containing the flaw

Equation C.26 specifies a threshold stress level as well as provides a flaw size towards failure (i.e., the residual strength of the pipe). In the corrosion of buried pipes, a corrosion pit present on the pipe can be considered as a notch in the pipe wall. If the corrosion pit is assumed to have a hemispherical shape, then the reduction in the pipe strength can be accounted for by modifying a methodology developed by Smith (1977) to determine the effect of a notch on the strength of a material. In the article, Smith showed that the relationship between the fracture toughness, material fatigue strength, and notch depth can be represented by:

$$\left(\left(1+2\sqrt{\frac{y}{r_n}}\right)\frac{\sigma_{nom}}{\sigma_0}\right)^2 - 1 + 4\frac{y}{r_n} = \frac{4}{\pi r_n}\left(\frac{K_0}{1.12\sigma_{nom}}\right)^2 \tag{C.27}$$

where: y = the depth of the notch, in

r_n = the radius of the root of the notch, in

σ_0 = the fatigue strength of the material, psi

σ_{nom} = the smallest stress level to cause complete failure, psi

K_0 = the fracture toughness of the material, psi$\sqrt{in}$

In the proposed model, the radius of the notch root is assumed to be equal to the notch depth. Equation C.27 then reduces the equation to

$$\left(3\frac{\sigma_{nom}}{\sigma_0}\right)^2 + 3 = \frac{4}{\pi r_n}\left(\frac{K_0}{1.12\sigma_{nom}}\right)^2 \tag{C.28}$$

Multiplying through by σ^2_{nom} leads to:

$$\sigma^4_{nom}\left(\frac{3}{\sigma_0}\right)^2 + 3\sigma^2_{nom} - \frac{4}{\pi r_n}\left(\frac{K_0}{1.12}\right)^2 = 0 \tag{C.29}$$

This is a quadratic equation in σ_{nom} and σ_{nom} is determined as:

$$\sigma_{nom} = \sqrt{\frac{\sigma^2_0}{6}\left(-1+\sqrt{\left[1+\frac{12.75}{\pi r_n}\left(\frac{K_0}{\sigma_0}\right)^2\right]}\right)} \tag{C.30}$$

Equation C.30 represents the smallest stress required to cause a failure in the pipe. This stress is taken as the residual strength of the pipe as a result of corrosion pits occurring in the walls.

A pipe's vulnerability is assumed to be controlled by the ratio of each of four major strength parameters to the stresses on the pipe. These strength parameters are tensile strength, bursting tensile strength, flexural modulus (beam strength), and ring modulus represented by σ_0 in Equation C.30. For each of these strengths, the above equation is used to compute the residual strength. The residual strength is then used in PBM to determine the pipe's vulnerability.

Pipe Break Module (PBM)

The PBM calculates a safety factor for each of the major stress components as the ratio of the residual strength of the pipe divided by the maximum expected stress. As the pipe ages and corrosion pits reduce the pipe wall thickness and concentrate stresses, the safety factor is reduced. Theoretically, the pipe will fail at a safety factor of 1 or less. In addition to the single stress components (ring, hoop, longitudinal, and beam stress) safety factors are computed for combined ring and hoop stress and for combined longitudinal and beam stress.

Generally, the safety factors are obtained by considering the six stress types: ring, flexural, tensile, hoop, combined hoop and ring, and combined flexural and longitudinal. In order to determine the safety factor of a pipe, the residual strength is estimated for the four strength categories (ring modulus, tensile strength, bursting tensile strength and modulus of rupture):

$$\sigma_{res} = \sqrt{\frac{\sigma_0^2}{6}\left(-1+\sqrt{\left[1+\frac{12.75}{\pi r_n}\left(\frac{K_0}{\sigma_0}\right)^2\right]}\right)} \qquad \text{(C.31)}$$

where r_n = the depth of the corrosion pit, in

σ_0 = the fatigue strength of the material, psi (taken as the original pipe strength for ring, tensile, moduli of rupture and flexure)

σ_{res} = the residual strength for each of strength category (ring modulus, tensile strength, bursting tensile strength, and moduli of rupture and flexure), psi

K_0 = the fracture toughness of the material, psi$\sqrt{\text{in}}$

Computation of Safety Factors

Using the residual strengths and the stresses computed from the PLM, the safety factors can be computed as follows:

Flexural safety factor. This is the safety factor resulting from bending (flexural) stresses. It represents the ratio of the residual modulus of rupture to the bending stresses. It is defined as:

$$SF_f = \frac{\sigma_{res(mr)}}{\sigma_f} \quad \text{(C.32)}$$

where SF_f = safety factor due to flexural stresses

$\sigma_{res(mr)}$ = residual modulus of rupture, psi.

σ_f = sum of flexural stresses, psi.

Ring safety factor. It represents the ratio of the residual ring modulus to the ring stresses. It is defined as:

$$SF_\theta = \frac{\sigma_{res(rm)}}{\sigma_\theta} \quad \text{(C.33)}$$

where SF_θ = safety factor due to ring stresses

$\sigma_{res(rm)}$ = residual ring modulus of rupture, psi.

σ_θ = sum of ring stresses, psi.

Hoop safety factor. It represents the ratio of the residual bursting tensile strength to the hoop stresses. It is defined as:

$$SF_h = \frac{\sigma_{res(bts)}}{\sigma_h} \quad \text{(C.34)}$$

where SF_h = safety factor due to hoop stresses

$\sigma_{res(bts)}$ = residual bursting tensile strength, psi.

σ_h = sum of hoop stresses, psi.

Longitudinal safety factor. It represents the ratio of the residual tensile strength to the longitudinal stresses. It is defined as:

$$SF_l = \frac{\sigma_{res(ts)}}{\sigma_l} \tag{C.35}$$

where SF_l = safety factor due to longitudinal stresses

$\sigma_{res(ts)}$ = residual tensile strength, psi.

σ_l = sum of longitudinal stresses, psi.

Flexural Plus Longitudinal Safety Factor. This safety factor represents the ratio of the residual modulus of rupture to the sum of flexural and longitudinal stresses. It is defined as:

$$SF_{flexlong} = \frac{\sigma_{res(ts)}}{\sigma_{flexlong}} \tag{C.36}$$

where $SF_{flexlong}$ = safety factor due to flexural plus longitudinal stresses

$\sigma_{res(ts)}$ = residual tensile strength, psi.

$\sigma_{flexlong}$ = sum of flexural and longitudinal stresses, psi.

Ring Plus Hoop Safety Factor. The safety factor for combined ring and hoop stress is determined by rewriting the combined load equation:

$$\frac{w^2}{W^2} = \frac{(P-p)}{P} \tag{C.37}$$

where: w = the external load, lb/ft

W = the crushing load, lb/ft

p = the internal pressure, psi

P = the bursting pressure, psi

Using the above equation the loads w and p are both replaced by F*w and F*p respectively. F is defined as the safety factor for combined hoop and ring stress. Solving for F yields:

$$F = \frac{-\left(\frac{p}{P}\right) + \sqrt{\left(\frac{p}{P}\right)^2 + 4\left(\frac{w}{W}\right)^2}}{2\left(\frac{w}{W}\right)^2} \tag{C.38}$$

Table C.3 is an index, in matrix form, showing which equations are used to calculate each component of stress on buried pipes. Tables C.4 and C.5 are examples of stress calculations for a buried pipe at year zero (new pipe) and at year 40 (aged pipe). It is clear that by ranking the SF in ascending order, the candidate pipes for replacement can be identified.

Table C.3

Index to equations for pipe stress calculations

	Load (lb/L.F.)	Pressure (psi)	Ring Stress (psi)	Hoop Stress (psi)	Flexural Stress (psi)	Longitudinal Stress (psi)	Flexural + Longitudinal Stress (psi)	Ring + Hoop Stress (psi)
Earth	C.1		C.12		C.20			
Traffic	C.3 & C.4		C.12		C.20			
Frost	C.8		C.12		C.20			
Internal pressure		Specified		C.15		C.21		
Thermal load on pipe		Specified temperature drop	C.13			C.22		
Water hammer		C.10		C.16		C.21		
Longitudinal component of hoop stress						C.23		
Expansion/contraction thermal stress						C.11		
Total stress	Sum		C.14			C.24		
Yield strength			C.31	C.31	C.31	C.31	C.31	
Safety factor			C.33	C.34	C.32	C.35	C.36	C.38

Table C.4

Summary of pipe load, stress, and safety factor calculations

	Load (lb/L.F.)	Pressure (psi)	Ring Stress (psi)	Hoop Stress (psi)	Flexural Stress (psi)	Longitudinal Stress (psi)	Flexural + Longitudinal Stress (psi)	Ring + Hoop Stress (psi)
Earth	1114.20		4431.70		1228.55			
Traffic	415.20		1651.40		457.80			
Frost	942.60		3749.08		1039.32			
Internal Pressure		100.00		975.61		487.80		
Thermal Load on Pipe		6.00	364.64			341.68		
Water Hammer		120.00		1170.73		585.37		
Longitudinal Component of Hoop Stress						450.73		
Expansion/Contraction Thermal Stress						3534.00		
Total Stress	2472.00		10196.83	2146.34	2725.67	5399.59	8125.26	
Yield Strength (F)			45000.00	21000.00	27426.00	30000.00	27426	
Safety Factor (F)			4.41	9.78	10.06	5.56	3.38	3.41

Conditions at Year 0

Pipe ID	5825014
Year	1961
Age (yrs)	0
Diameter (in)	8
Length (ft)	180.4
Pit depth (in)	0.000
Wall thickness (in)	0.410

Table C.5

Summary of pipe stress calculations at year 40

	Load (lb/L.F.)	Pressure (psi)	Ring Stress (psi)	Hoop Stress (psi)	Flexural Stress (psi)	Longitudinal Stress (psi)	Flexural + Longitudinal Stress (psi)	Ring + Hoop Stress (psi)
Earth	1114.20		7071.25		1579.05			
Traffic	415.20		2634.98		588.41			
Frost	942.60		5982.05		1335.83			
Internal Pressure		100.00		1238.80		619.40		
Thermal Load on Pipe		6.00	362.30			344.03		
Water Hammer		120.00		1486.55		743.28		
Longitudinal Component of Hoop Stress						572.32		
Expansion/Contraction Thermal Stress						3534.00		
Total Stress	2472.00		16050.58	2725.35	3503.29	5813.03	9316.32	
Yield Strength (F)			19100.57	14939.22	16443.90	16945.70	16443.9	
Safety Factor (F)			1.19	5.48	4.69	2.92	1.77	0.92

Conditions at Year 40

Pipe ID	5825014
Year	2001
Age (yrs)	40
Diameter (in)	8
Length (ft)	180.4
Pit depth (in)	0.087
Wall thickness (in)	0.323

Data Requirements

The data requirements for the modules described in the previous sections fall into three main categories, pipe data, soil data, environmental data, and regional weather data. The data requirements are summarized in Table C.6.

Table C.6

Required input data

Parameter	Description
Pipe material	Currently only cast iron pipes are considered in this model
Date of installation, year	Determines manufacturing method, initial strength of the pipe, and standard pipe dimensions
Pipe diameter, in.	Stress calculation
Pipe thickness	Standard pipe size can be used if not known.
Depth of pipe	Affects the earth load and traffic load on pipe
Soil unit weight, lb/ft^3	Affects the earth load on pipe
Soil type, USGS standard classification	Affects earth and traffic load on pipe and corrosivity of soil
Soil pH	Affects corrosivity of soil
Soil Resistivity, ohm/cm	Affects corrosivity of soil Lower resistivity values increase corrosivity
Soil moisture content, %	Affects frost load calculation
Is pipe under a street?, y/n	Determines whether traffic loads are calculated
Is street paved?, y/n	Affects traffic load calculation
Truck load, heavy or light	Affects traffic load calculation
Minimum and maximum water temperature, °F	Affects thermal stress calculations
Maximum temperature drop, °F/day	Affects thermal stress calculations
Number of consecutive days when air temperature remains below freezing	Affects frost load calculation
Maximum frost depth, in.	Affects frost load calculation

PROPERTIES OF GRAY CAST IRON PIPE

Various manufacturing techniques have been used to make gray cast iron pipes at different times. Therefore, the date of manufacture, date of installation, or age of a pipe can be used to estimate the manufacturing technique used to make the pipe. The manufacturing techniques directly affect the inherent material properties of the pipes.

Prior to 1850, gray cast iron pipes were manufactured using a horizontal casting technique (O'Day et al. 1986). Horizontal cast pipes were cast in molds in lengths of four or five ft. These pipes were characterized by short lengths and uneven wall thickness that could include inclusions of sand or slag, lowering the strength of the pipe material.

Between 1850 and the 1930s, gray iron pipes were pit cast in vertical molds in lengths of 12 ft. These pipes were more uniform in thickness than the horizontal cast pipes with fewer inclusions. The longer pipe lengths reduced the number of joints for a given length of pipe.

Between the 1930s and 1950s, gray iron pipes were centrifugally cast (spun cast). This technique allowed the pipe to be thinner, with a more uniform pipe wall. In addition, improvements in the composition of the metal resulted in stronger pipe material.

The following summary Tables C.7 and C.8 of pipe properties inherent in pipes manufactured with different manufacturing techniques were adapted from PWD (1985a and 1985b).

Table C.7

Cast iron strength measurements

Classification	PWD use	Bursting tensile strength, psi	Ring modulus of rupture, psi	Tensile strength, psi
Horizontal cast	Pre-1850	Not Available	Not Available	Not Available
Pit cast	1850-1930s	11,000	31,000	20,000
Centrifugally cast	1930s	18,000	40,000	35,000
Centrifugally cast	1950s	21,000	45,000	35,000

Source: PWD 1985a.

Table C.8

Estimation of design thickness

Period	6-in.	8-in.	10-in.	12-in.
Before 1901	0.47	0.47	0.58	0.63
1901-1908	0.44	0.44	0.5	0.56
1909-1938	0.43	0.46	0.5	0.54
1939-1951	0.28	0.32	0.37	0.38
1952-1966	0.38	0.41	0.44	0.48

Source: PWD 1985b.

The information from PWD was supplemented with lab tests performed on pipe samples collected as part of this project. Tables C.9 through C.14 summarize the manufacturing dates and material properties used in the PBM.

Table C.9

Fracture toughness summary computed from NRC data

Pipe material	Average (psi*in.^{2})	Standard deviation (psi*in.^{2})	Minimum (psi*in.^{2})	Maximum (psi*in.^{2})
spun cast	12032	1409	9365	13993
pit cast	10693	1585	5202	12426

Source: Summarized from Rajani et al. 2000.

Table C.10

Summary of fracture toughness test results (Lehigh Labs) conducted in this project

Pipe ID	Pipe type	Installed	Fracture toughness ($psi*in.^2$)
PSW07	Spun cast	1948	91478
PSW14	Pit cast	1923	70117
PWB05	Spun cast	1960	82179
PWB06	Spun cast		75948
PWB07	Spun cast	1953	79809
PWB08	Spun cast		90953
UWNJ01	Spun cast		62644
UWNJ04	Spun cast		16211
UWNJ06		1921	55849
Average			69465
Standard deviation			23252
Minimum			16211
Maximum			91478

Table C.11

Summary of fracture toughness test results (NRC Labs) conducted in this project

Pipe ID	Pipe type	Installed	Fracture toughness ($psi*in.^2$)
RMOC01	Spun cast	1961	10374
RMOC01	Spun cast	1961	10374
RMOC02	Spun cast	1962	10738
RMOC04	Spun cast	1954	10920
RMOC05	Spun cast	1954	10101
RMOC06	Spun cast	1960	11102
RMOC08	Spun cast	1960	11102
RMOC09	Spun cast	1977	11830
RMOC10	Spun cast	1971	12558
RMOC15	Spun cast	1958	12922
RMOC16	Spun cast	1961	10829
RMOC17	Spun cast	1961	10192
RMOC18	Spun cast	1955	11830
RMOC19	Spun cast	1963	10829
Average	Spun cast	1963	11179
Standard deviation	Spun cast	1963	870
Minimum	Spun cast	1963	10101
Maximum	Spun cast	1963	12922

Table C.12

Summary of modulus of rupture test results (NRC Labs) conducted in this project

Pipe ID	Pipe type	Pipe size (in)	Installed	Mod Rup (psi)
RMOC15	Spun cast	6	1958	51246
RMOC18	Spun cast	6	1955	55626
RMOC03	Spun cast	4	1950	42194
RMOC10	Spun cast	8	1971	53436
RMOC05	Spun cast	6	1954	34748
RMOC01	Spun cast	6	1961	32266
RMOC02	Spun cast	8	1962	48034
RMOC09	Spun cast	6	1977	48764
RMOC17	Spun cast	6	1961	42194
RMOC16	Spun cast	6	1961	48180
RMOC08	Spun cast	8	1960	14016
RMOC06	Spun cast	8	1960	52268
RMOC04	Spun cast	16	1954	48618
Average (psi)				43968
Standard deviation (psi)				11371
Minimum (psi)				14016
Maximum (psi)				55626

Table C.13

Summary of modulus of rupture test results (Lehigh Labs) conducted in this project

Pipe ID	Pipe type	Pipe size (in)	Installed	Mod Rup (psi)
PSW14	Pit cast		1923	11415
PSW06	Spun cast	6	1950	9815
PSW07	Spun cast	6	1948	23589
PSW08	Spun cast	8	1951	18618
PSW12	Spun cast	6	1950	9815
PWB05	Spun cast	6	1960	18908
PWB06	Spun cast	4		9662
PWB07	Spun cast	8	1953	10835
PWB08	Spun cast	4		21324
UWNJ01	Spun cast	6		8672
UWNJ02	Spun cast	6		10119
UWNJ04	Spun cast	6		6768
UWNJ05	Spun cast	6		8538
UWNJ06		8	1921	8503
UWNJ07		6	1961	19776
Average (psi)				13090
Standard deviation (psi)				5595
Minimum (psi)				6768
Maximum (psi)				23589

Table C.14

Outside diameters of cast iron pipe

Pipe size (in)	Outside diameter (in)
3	3.96
4	4.80
6	6.90
8	9.05
10	11.10
12	13.20
14	15.30
16	17.40
18	19.50

Source: CIPRA 1978.

APPENDIX D
MECHANISTIC MODEL USERS GUIDE

PIPEADDIN SOFTWARE OVERVIEW AND SUMMARY OF FEATURES

The PIPEADDIN software was designed for use by water utilities for analyzing and prioritizing their cast iron water main pipe replacement needs. The software primarily focuses on small diameter pipes (4-12 inches) and is limited to cast iron pipes. The software allows the user to provide basic information related to the pipe or pipes to be analyzed and then uses the information provided to model the deterioration of the pipe over time and also to model the loads and stresses on the pipe. The results are used to compute the residual strength SF for each pipe in the analysis. Using the resulting estimated residual SF the user can then rank the pipes in increasing order of SF to develop a prioritized replacement plan.

The analysis tasks are accomplished using three main models Pipe Deterioration Module (PDM), Pipe Load Module (PLM), and Pipe Break Module (PBM). The software was developed as an ADD-IN that runs in Microsoft Excel. The Excel environment was chosen to facilitate ease of use of the software.

The software is composed of three Excel workbook files. The first file (pipeaddin.xls) contains the macros and functions that make up the pipe analysis program. The second file which is referred to in this document as the parameters file is made up of sheets that contain default design values for the fracture toughness, modulus of rupture, ring modulus, tensile strength, bursting tensile strength, and design thickness. These sheets will be discussed again later in this document. The third workbook, referred to as the data workbook, contains the data for the pipes to be analyzed. Both the data workbook and the parameters workbook may be saved under a different name but the data format must remain consistent with the default versions of these workbooks.

System Requirements

Design System Configuration:

The software was developed on a Pentium II 400 computer with the following configuration:

RAM: 128 MB RAM

OS: Microsoft Windows NT Workstation 4.0.

STORAGE: 8.4 GB

SOFTWARE: Microsoft Excel 97

The program has been tested successfully on Microsoft Windows 95, 98, 2000, and Windows NT 4.0. It has also been successfully tested on Microsoft Excel 97 and 2000. Typically, about 50 MB hard drive space is needed for the data preprocessing.

Recommended Minimum System Requirements:

Minimum system requirements for running Microsoft Office 97

64 MB of RAM (More memory improves performance)

Installation

Important Installation Notes

In order to run the program,

1. Make sure you have *Microsoft Office (97 or later required)* installed
2. In Excel, make sure that the *Analysis ToolPak* and *Analysis ToolPak-VBA* addins are installed and activated.

If you are not sure whether you have the addins installed, do the following:

a. in the Excel menu, click 'Tools'

b. Select 'Add-ins..' This should give you a list of all the available Addins make sure that the *Analysis ToolPak* and *Analysis ToolPak-VBA* addins are checked. If they are not checked, place a check in the box next to *Analysis ToolPak* and *Analysis ToolPak-VBA*, then click *ok* to make these features available.

Required files

The main program consists of the following files: the program file (**pipeaddin.xls**), the data file (**data.orig.xls**) and the parameters file (**param.orig.xls**). All three files must be open in order for the program to run properly.

The program file (**pipeaddin.xls**):

The file contains all the macros for the software. No changes should ever be made to this file.

The data file (**data.orig.xls**):

This file contains data on pipe and related properties required for analysis. Information in this file may be changed and the file may be renamed but the renamed file must start with 'data.'. So files name data.test.xls, data.new.xls, and data.monday.xls are all acceptable. Note that the original sample file packaged with the software is named data.orig.xls. This file contains an example pipe break database, soil definitions, and region definitions.

The data file (**param.orig.xls**):

This file contains data on basic settings to used in the analysis. It also contains a settings sheet that allows the user to specify how the program computes certain parameters. Information in the file may be changed and the file may be renamed but the renamed file must start with 'param.'. So files name param.test.xls, param.new.xls, and param.monday.xls are all acceptable file names. Note that the original sample file packaged with the software is named param.orig.xls. This file contains default parameter settings for computation methods and material properties such as bursting tensile strength, ring modulus, and modulus of rupture.

Installation procedure

Select a directory to install the program in. Copy the pipeaddin.xls file to that directory. The data file (example data.orig.xls) and the parameters file (example param.orig.xls) may reside in the same folder or in a different folder.

Running The Program

The following is how the program can be started:

1. Load the file pipeaddin.xls. If prompted to enable macros, **choose to enable macros**. (This is required to run the program)
2. Once this file is loaded, you should see a menu item, "PipeADDIN", added to the Excel menu, as shown in Figure D.1.

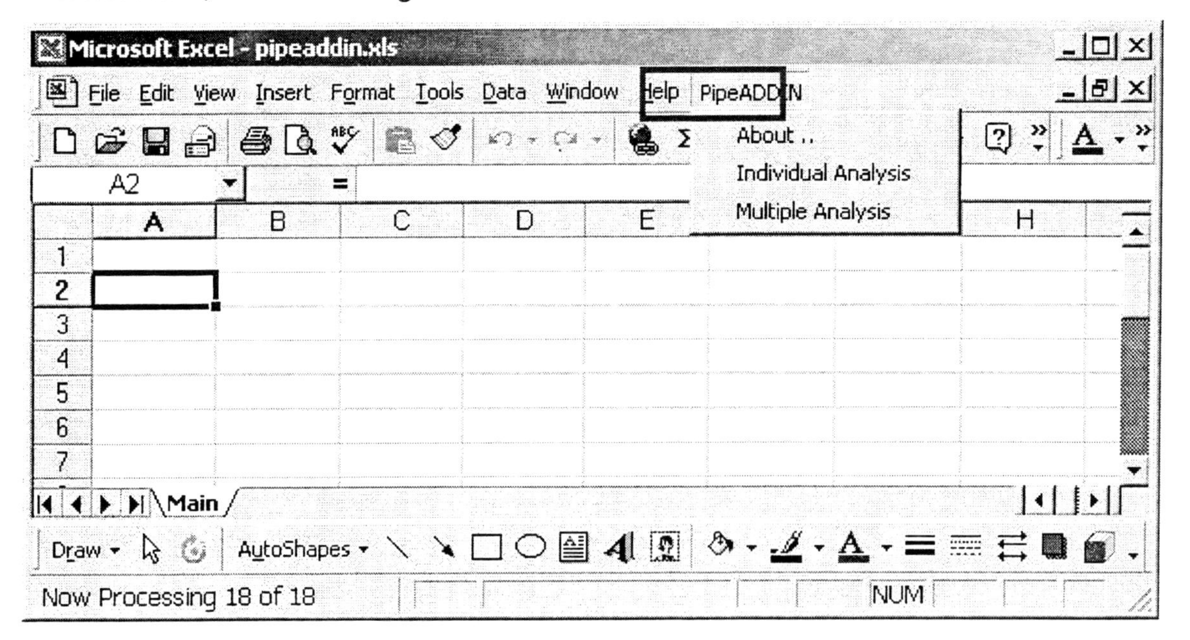

Figure D.1 Main program sheet

3. Load the data.orig.xls file or the appropriate data file that you would like to use. Also, load the param.orig.xls file or the appropriate parameters file.

   Note: Please make sure the sheets in the data file and the parameters files are in the same format those in the data.orig.xls and param.orig.xls provided with the software.

Required Data and Data Input

In order the input data for analysis, it is recommended that the user edit the sample file (data.orig.xls) provided with the software. This section describes the critical and optional data needs. Data for the software is input in three main sheets in the data workbook. These sheets are *BrkInventory, Soils*, and *Regions.* All the other sheets are used to display results and they will be discussed later.

The *BrkInventory* Sheet

This sheet (see Figure D.2) contains data related to each pipe to be analyzed. Table D.1 contains each field, its description and whether or not it is required. This sheet also allows specific information to be provided for each pipe.

data.orig.xls

	A	B	C	D	E	F	G	H	I	
1	PIPEID	RegionID	SoilClass	AnalysisYear	PipeType	PipeDiameter	PipeYearInstalled	PipeLength	TrafficType	Wor
2	5825014	RMOC	7	1997	spuncast	8	1961	180.4	HEAVY	
3	6021130	RMOC	4	1992	spuncast	6	1966	770.8	LIGHT	
4	6224067	RMOC	4	1991	spuncast	6	1958	820	NONE	
5	6226013	RMOC	2	1996	spuncast	6	1954	787.2	MEDIUM	
6	6226171	RMOC	2	1981	spuncast	6	1958	524.8	HEAVY	
7	6426099	RMOC	2	1981	spuncast	6	1952	551.04	MEDIUM	
8	6428276	RMOC	2	1993	spuncast	6	1939	531.36	HEAVY	

BrkInventory / Soils / Regions / Summary / ResultsIndividualPipe / ResultsAtZeroYea

Figure D.2 Pipe inventory sheet

Table D.1

List of parameters in Inventory sheet

Parameter	Description	Required
PIPEID	This is the unique identification for the pipe.	y
RegionID	This is an ID cross listed in the 'Regions' sheet used to specify environment factors affecting the pipe.	y
SoilClass	This is an ID cross listed in the 'Soils' sheet used to specify soil conditions surrounding the pipe.	y
AnalysisYear	This is the year in which the pipe is being analyzed.	y
PipeType	spun cast or pit cast	y
PipeDiameter	Pipe size, inches	y
PipeYearInstalled	Year in which pipe was installed	y
PipeLength	Length of pipe, ft	n
TrafficType	Traffic volume if applicable, heavy, medium, light, none	y
WorkingPressure	The pressure within the pipe, psi	y
PipeDepth	Depth at which the pipe is buried, ft	y
PavementType	Whether street is paved or unpaved	y
BeamSpan	Estimated unsupported beam span length, ft	y
SoilType	Soil type, this can be provided if the soil around this pipe has been sampled. Value entered here over rides the type for the soil class on the 'Soils' sheet	n
SoilpH	Soil pH, this can be provided if the soil around this pipe has been sampled. Value entered here over rides the pH for the soil class on the 'Soils' sheet	n
SoilResistivity	Soil resistivity (ohm-cm), this can be provided if the soil around this pipe has been sampled. Value entered here over rides the resistivity for the soil class on the 'Soils' sheet	
SoilAeration	Soil aeration (good, fair, poor), this can be provided if the soil around this pipe has been sampled. Value entered here over rides the aeration for the soil class on the 'Soils' sheet	n

(continued)

Table D.1 (continued)

Parameter	Description	Required
SoilMoisture	Soil moisture content, this can be provided if the soil around this pipe has been sampled. Value entered here over rides the moisture content for the soil class on the 'Soils' sheet. Soil moisture is used in the computation of the frost load provided the soil is frost susceptible. In relation to the moisture content, the worst case frost load occurs when the soil is saturated. Hence the saturated moisture content will provide the maximum frost load.	N
SoilDensity	Soil density (lb/ft^3), this can be provided if the soil around this pipe has been sampled. Value entered here over rides the density for the soil class on the 'Soils' sheet	n
SoilPorosity	Soil porosity (%), this can be provided if the soil around this pipe has been sampled. Value entered here over rides the porosity for the soil class on the 'Soils' sheet	n
FrostSusceptible	Frost susceptible (y, n) This states whether the soil is expansive or not. If a 'n' is selected, no frost loads will be computed. Value entered here over rides the frost susceptibility for the soil class on the 'Soils' sheet	n
CorrosionRate	Corrosion Rate (in/yr) If a corrosion is specified, it used to compute the pit depth over time. Otherwise, the Adjusted Rossum model is used to compute the pit depth.	N
TensileStrength	Tensile Strength, psi. This is the tensile strength of the pipe at the time it installed. Value entered here overrides the tensile strength on the 'TensileStrength' sheet in the pipeaddin.xls workbook	n

(continued)

Table D.1 (continued)

Parameter	Description	Required
BurstingTensileStrength	Bursting Tensile Strength, psi. This is the bursting tensile strength of the pipe at the time it installed. Value entered here overrides the bursting tensile strength on the 'BurstingTensileStrength' sheet in the pipeaddin.xls workbook	n
RingModulus	Ring Modulus, psi. This is the ring modulus of the pipe at the time it installed. Value entered here overrides the ring modulus strength on the 'RingModulus' sheet in the pipeaddin.xls workbook	n
ModulusRupture	Modulus of Rupture, psi. This is the modulus of rupture of the pipe at the time it installed. Value entered here overrides the modulus of rupture on the 'ModulusRupture' sheet in the pipeaddin.xls workbook	n
FractureToughness	Fracture Toughness, psi. This is the fracture toughness of the pipe at the time it installed. Value entered here overrides the fracture toughness on the 'FractureToughness' sheet in the pipeaddin.xls workbook	n
DesignThickness	Design thickness, inches. This is the thickness of the pipe at the time it installed. Value entered here overrides the design thickness on the 'DesignThickness' sheet in the pipeaddin.xls workbook.	n

The *Soils* Sheet

This sheet (see Figure D.3) contains data related to the various soil classifications used in the pipe analysis. Table D.2 contains the description of the soil parameters and whether or not they are required. This table is useful for grouping soils into general categories or classes and assigning each pipe to these soil classes by designating the 'SoilClass' on the BrkInventory sheet.

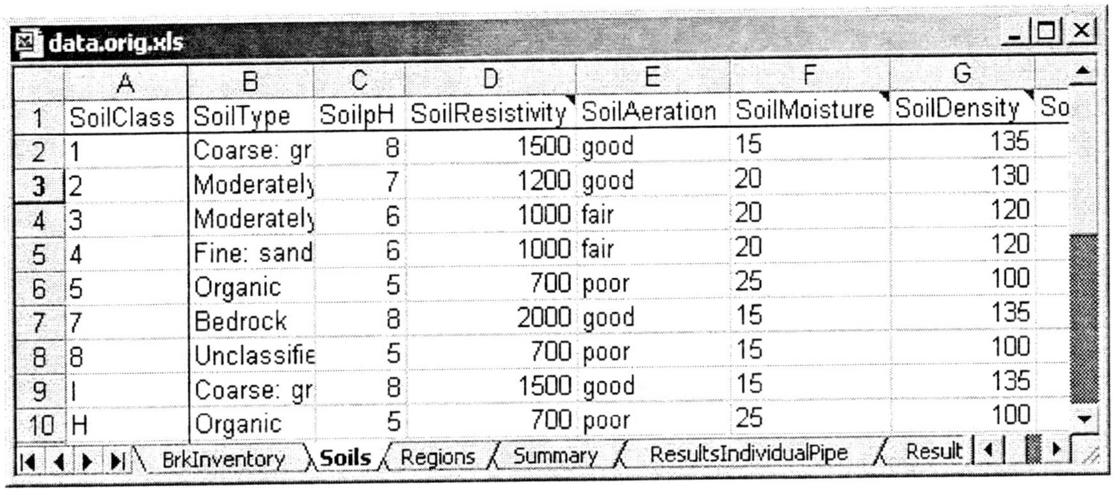

data.orig.xls

	A	B	C	D	E	F	G	
1	SoilClass	SoilType	SoilpH	SoilResistivity	SoilAeration	SoilMoisture	SoilDensity	So
2	1	Coarse: gr	8	1500	good	15	135	
3	2	Moderatel)	7	1200	good	20	130	
4	3	Moderatel)	6	1000	fair	20	120	
5	4	Fine: sand	6	1000	fair	20	120	
6	5	Organic	5	700	poor	25	100	
7	7	Bedrock	8	2000	good	15	135	
8	8	Unclassifie	5	700	poor	15	100	
9	I	Coarse: gr	8	1500	good	15	135	
10	H	Organic	5	700	poor	25	100	

BrkInventory / Soils / Regions / Summary / ResultsIndividualPipe / Result

Figure D.3 Soils sheet.

Table D.2

List of parameters in Soils sheet

Parameter	Description	Required
SoilClass	This is an ID cross listed in the 'BrkInventory' sheet and is used to link a pipe to its related soil parameters.	y
SoilType	Soil type	y
SoilpH	Soil pH, this is needed in computing the corrosion pit depth over time.	y
SoilResistivity	Soil resistivity (ohm-cm) , this is needed in computing the corrosion pit depth over time.	y
SoilAeration	Soil aeration (good, fair, poor) , this is needed in computing the corrosion pit depth over time.	y
SoilMoisture	Soil moisture content (%) Soil moisture is used in the computation of the frost load provided the soil is frost susceptible. In relation to the moisture content, the worst case frost load occurs when the soil is saturated. Hence the saturated moisture content will provide the maximum frost load.	y
SoilDensity	Soil density (lb/ft^3)	y
SoilPorosity	Soil porosity (%)	y
FrostSusceptible	Frost susceptible (y, n) This states whether the soil is susceptible to frost or not. If a 'n' is selected, no frost loads will be computed.	y

The *Regions* Sheet

This sheet (see Figure D.4) contains data related to the environmental and regional parameters related to the pipes. Table D.3 contains the description of the region parameters and whether or not they are required.

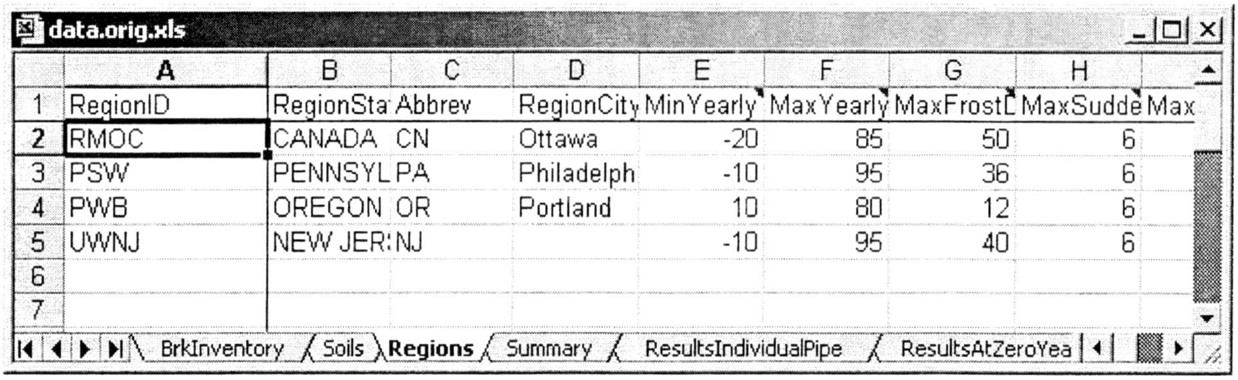

data.orig.xls

	A	B	C	D	E	F	G	H	
1	RegionID	RegionSta	Abbrev	RegionCity	MinYearly	MaxYearly	MaxFrostD	MaxSudde	Max
2	RMOC	CANADA	CN	Ottawa	-20	85	50	6	
3	PSW	PENNSYL	PA	Philadelph	-10	95	36	6	
4	PWB	OREGON	OR	Portland	10	80	12	6	
5	UWNJ	NEW JER:	NJ		-10	95	40	6	
6									
7									

BrkInventory / Soils / Regions / Summary / ResultsIndividualPipe / ResultsAtZeroYea

Figure D.4 Regions sheet.

Table D.3

List of parameters in Regions sheet

Parameter	Description	Required
RegionID	Unique Identifier for the region for which analysis is being performed	y
RegionState	Further defines the region by identifying the state or province.	n
Abbrev	Abbreviation for the state or province	n
RegionCity	Specific city within the region	n
MinYearlyTemp	Minimum temperature recorded in a year, F	y
MaxYearlyTemp	Maximum temperature recorded in a year, F	y
MaxFrostDepth	Maximum frost depth recorded, inches	y
MaxSuddenWaterTempChange	Maximum expect sudden drop or increase in water temperature, F	y
MaxFreezeDays	Maximum number of consecutive days below freezing, days	y
MinWaterTemp	Minimum water temperature per year, F	y
MaxWaterTemp	Maximum water temperature per year, F	y
WaterVelocityChange	Expected sudden water velocity change due to value shut down or change in pumping operations, ft/s	y

Default Design Parameters

The sheets shown in Figure D.5 to Figure D.11 are contained within the parameters file. These sheets contain estimated design parameters for pipes based on size and year of installation. They may be edited by the user.

The Settings Sheet

The settings sheet (Figure D.5) provides the user with the opportunity to change various options related to how the software computes the frost load and surge pressure. *For computing the frost load*, selecting '1' indicates the model will compute the frost load as an estimate of twice the computed earthload. Selecting '2' indicated the frost load will be computed based on the frost Rajani-Zahn load model. Either of these options may also be overwritten by the user providing the estimated frost load when running the software in individual pipe mode.

For the *surge pressure computation*, selecting '1' indicates the use of CIPRA recommended surge pressure values based on pipe size. Selecting '2' indicates the surge pressure will be computed based on the water hammer formula.

For the *target SF*, the user must select the minimum SF below which pipe must be targeted for replacement. This value must be greater than zero.

For the *Show All Pipes* Option, selecting '1' indicates that results for all pipes will be printed whether or not they are targeted for replacement. Selecting '0' indicates that only pipes targeted for replacement, that is only pipes with their minimum SF less than the target SF will be printed.

For the *Sort Results* Option, selecting '1' indicates that the results will be sorted on the "ResultsAtAnalysisYear" Sheet when running the software in Multiple Pipe mode. Selecting '0' indicates the results will not be sorted.

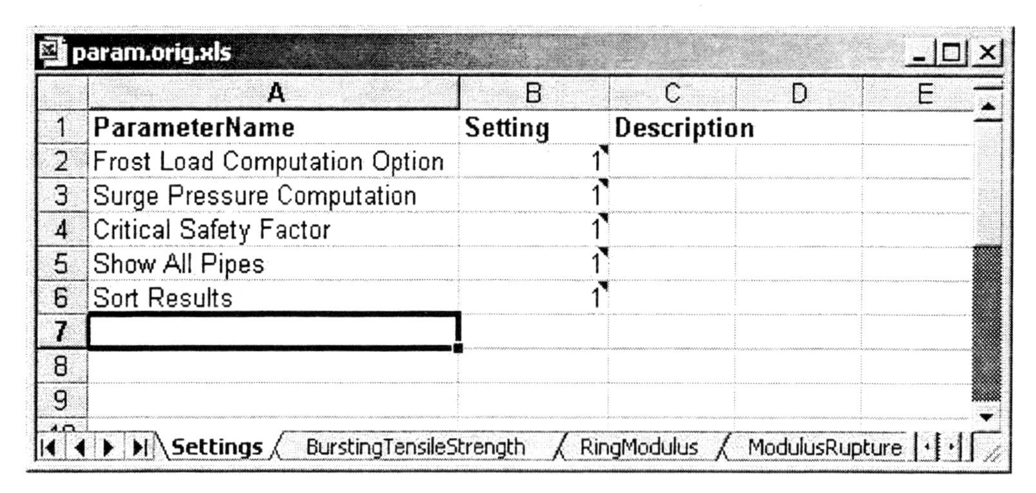

param.orig.xls

	A	B	C	D	E
1	ParameterName	Setting	Description		
2	Frost Load Computation Option	1			
3	Surge Pressure Computation	1			
4	Critical Safety Factor	1			
5	Show All Pipes	1			
6	Sort Results	1			
7					
8					
9					

Settings / BurstingTensileStrength / RingModulus / ModulusRupture

Figure D.5 Setting options

Bursting Tensile Strength

The bursting tensile strength (Figure D.6) is the amount of internal pressure that the pipe can withstand in the absence of external loads. The values may be obtained from pipe manufacture practices at the time of installation of the pipe.

param.orig.xls

	A	B	C	D	E	F	G	H	I
1	Year	Value	Notes						
5	1803	11000							
6	1804	11000							
7	1805	11000							
8	1806	11000							
9	1807	11000							
10	1808	11000							

Settings / BurstingTensileStrength / RingModulus / ModulusRupture / TensileStrength

Figure D.6 Default bursting tensile strength values

Ring Modulus

The ring modulus (Figure D.7) is the maximum stress the pipe can withstand before failure when subjected to external crushing load. It may be estimated based on design practices at the time of installation of the pipe.

param.orig.xls

	A	B	C	D	E	F	G	H	I
1	Year	Value	Notes						
5	1803	31000							
6	1804	31000							
7	1805	31000							
8	1806	31000							
9	1807	31000							
10	1808	31000							

Settings / BurstingTensileStrength / **RingModulus** / ModulusRupture / TensileStrength

Figure D.7 Default ring modulus values

Modulus of Rupture

The modulus of rupture (Figure D.8) is the bending strength of the pipe. The values may be obtained from pipe manufacture practices at the time of installation of the pipe or from tests on pipe samples.

param.orig.xls

	A	B	C	D	E	F	G	H	I
1	Year	Value	Notes						
167	1965	27426							
168	1966	27426							
169	1967	27426							
170	1968	27426							
171	1969	27426							
172	1970	27426							

Settings / BurstingTensileStrength / RingModulus / **ModulusRupture** / TensileStrength

Figure D.8 Default modulus of rupture values

Tensile Strength

The tensile strength (Figure D.9) is maximum tension stress the pipe can withstand before failure. The values may be obtained from pipe manufacture practices at the time of installation of the pipe.

param.orig.xls

	A	B	C	D	E	F	G	H	I
1	Year	Value	Notes						
145	1943	20000							
146	1944	20000							
147	1945	20000							
148	1946	20000							
149	1947	20000							
150	1948	20000							

Settings / BurstingTensileStrength / RingModulus / ModulusRupture / **TensileStrength** /

Figure D.9 Default tensile strength values

Fracture Toughness

Fracture toughness (Figure D.10) is the resistance of a material to failure from fracture starting from a preexisting crack. Values may be obtained from facture toughness tests.

param.orig.xls

	A	B	C	D	E	F	G	H	I
1	Year	Value	Notes						
160	1958	12032							
161	1959	12032							
162	1960	12032							
163	1961	12032							
164	1962	12032							
165	1963	12032							

RingModulus / ModulusRupture / TensileStrength / **FractureToughness** / DesignThickness

Figure D.10 Default fracture toughness values

Design Thickness

The design thickness (Figure D.11) is the original pipe wall thickness at installation. It may be determined by measurement from pipe sample, from pipe installation records, or based on design practices at time of installation.

param.orig.xls

	A	B	C	D	E	F	G	H	I
1	Year	4	6	8	10	12	14	Notes	
5	1803	0.47	0.47	0.47	0.58	0.63	0.63		
6	1804	0.47	0.47	0.47	0.58	0.63	0.63		
7	1805	0.47	0.47	0.47	0.58	0.63	0.63		
8	1806	0.47	0.47	0.47	0.58	0.63	0.63		
9	1807	0.47	0.47	0.47	0.58	0.63	0.63		
10	1808	0.47	0.47	0.47	0.58	0.63	0.63		

ModulusRupture / TensileStrength / FractureToughness / DesignThickness

Figure D.11 Default design thickness values

Using The Program

All the actions needed to perform the analysis are accomplished through the menu items in the drop down menu in Figure D.12.

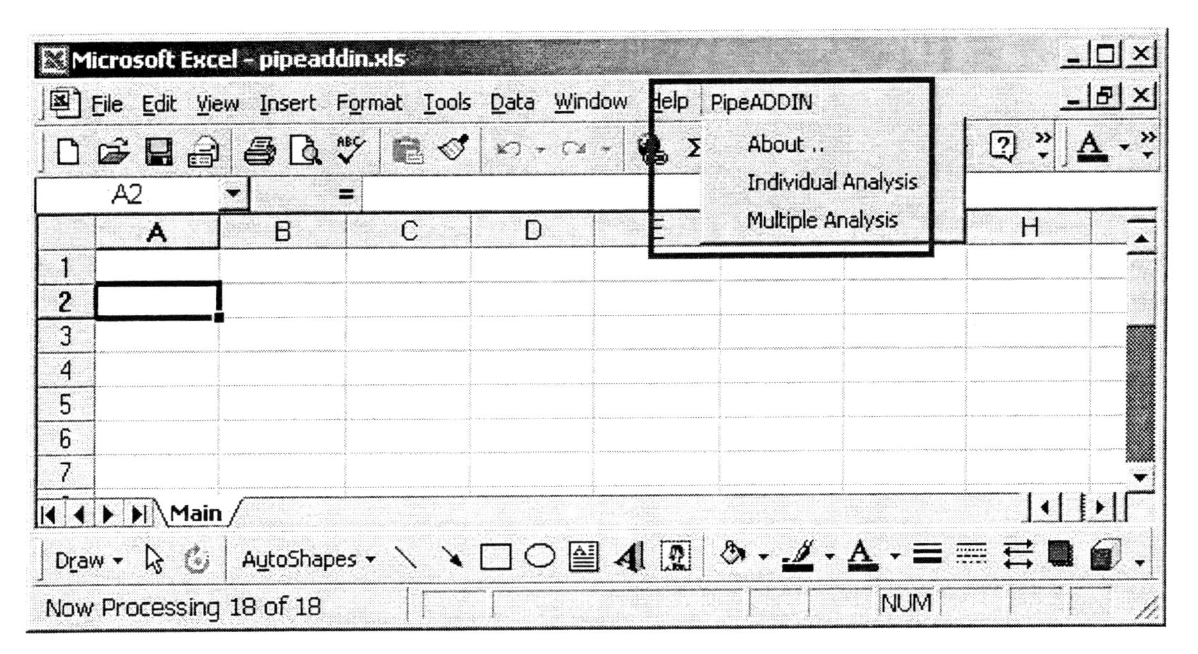

Figure D.12 Analysis options

The software permits the analysis of pipes in two ways: 1. Pipes could be analyzed individually permitting the user to alter both original and derived data for the individual pipe and observe how the pipe condition changes in response to the various scenarios. 2. Pipes can also be analyzed in batch mode allowing numerous pipes in a water distribution system to be analyzed at once.

Single Pipe Mode

In the single pipe mode, the software can be used to conduct detailed analysis on a pipe. To use the software in single pipe mode, select '*Individual Analysis*' from the '*PipeADDIN*' Menu (see Figure D.12). This brings up the form shown in Figure D.13. The user can then select a pipe to be analyzed from the drop down menu at the top. Whenever a pipe is selected, the properties related to the pipe are automatically read from the BrkInventory, Soils, and Regions sheet and the corresponding fields are updated in Figure D.13 to Figure D.16. The user can change any of the editable fields in Figure D.13 to Figure D.16 and then run the pipe analysis program. In this mode, the user can also chose to bypass the load models and specify specific estimated loads (see Figure D.15). Once the user has specified the parameters for the pipe, the button 'Analyze and Display Results' can then be clicked from within the "Report" tab and the various results are printed as shown in Figure D.17. In addition to the results printed in Figure D.17, the software also prints a formatted summarized report on the sheet 'Summary' in the data workbook. The summarized report sample is shown in Figure D.18 and Figure D.19. Figure D.18 contains the results for the pipe at the time it was installed and Figure D.19 is a summary of the current condition of the pipe. Finally, a year by year condition of the pipe (from the installation year to the analysis year) is printed on the sheet 'ResultsIndividualPipe' (see Figure D.20).

Pipe Properties Form

The default values for each pipe on this form are loaded from the 'BrkInventory' sheet when the pipe ID changes. The user can then change a specific value before continuing with the analysis.

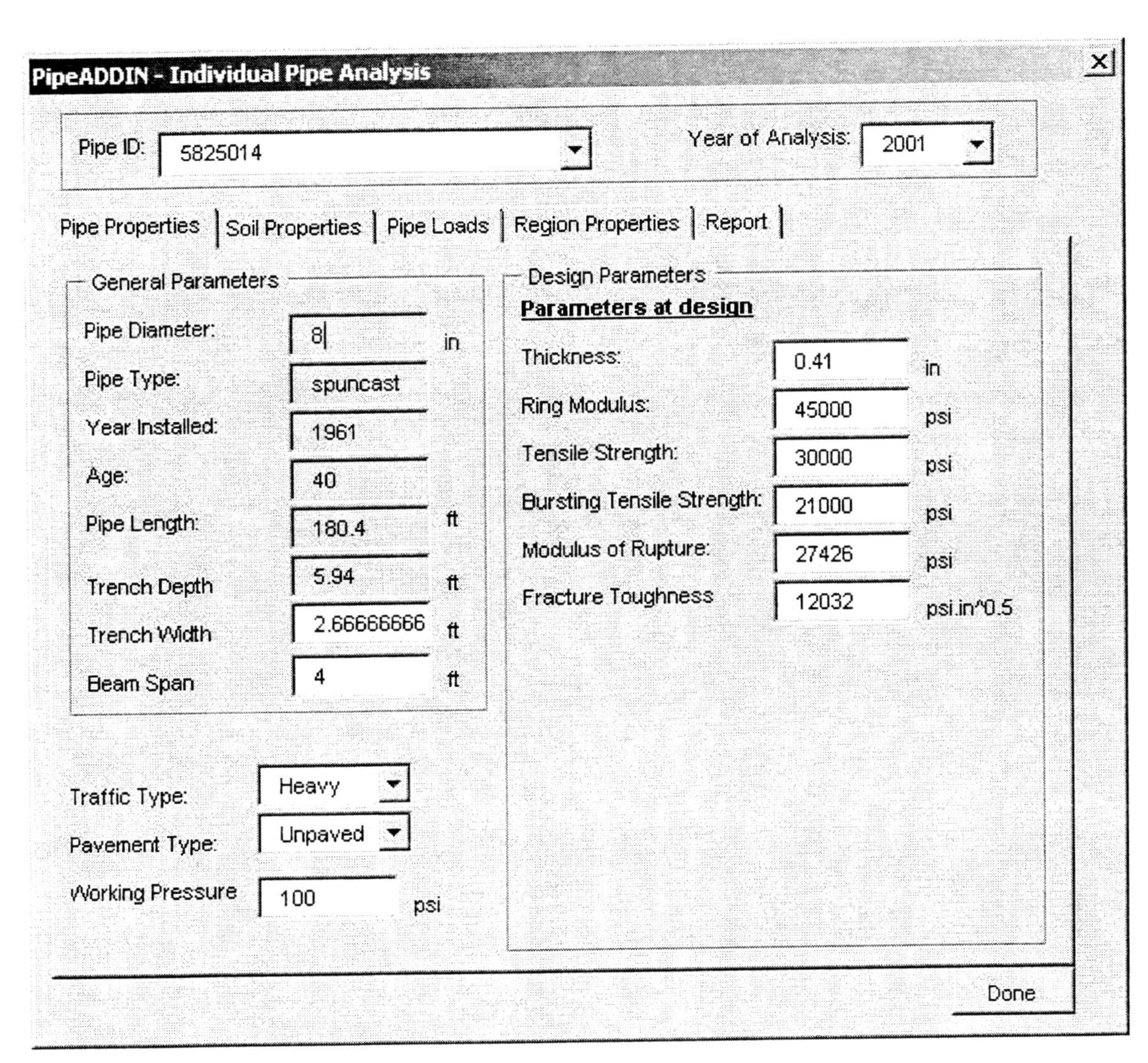

Figure D.13 Pipe properties for individual pipe mode

Soil Properties Form

The default values for each pipe on this form are loaded from the 'Soils' sheet when the pipe ID changes. The user can then change a specific value before continuing with the analysis.

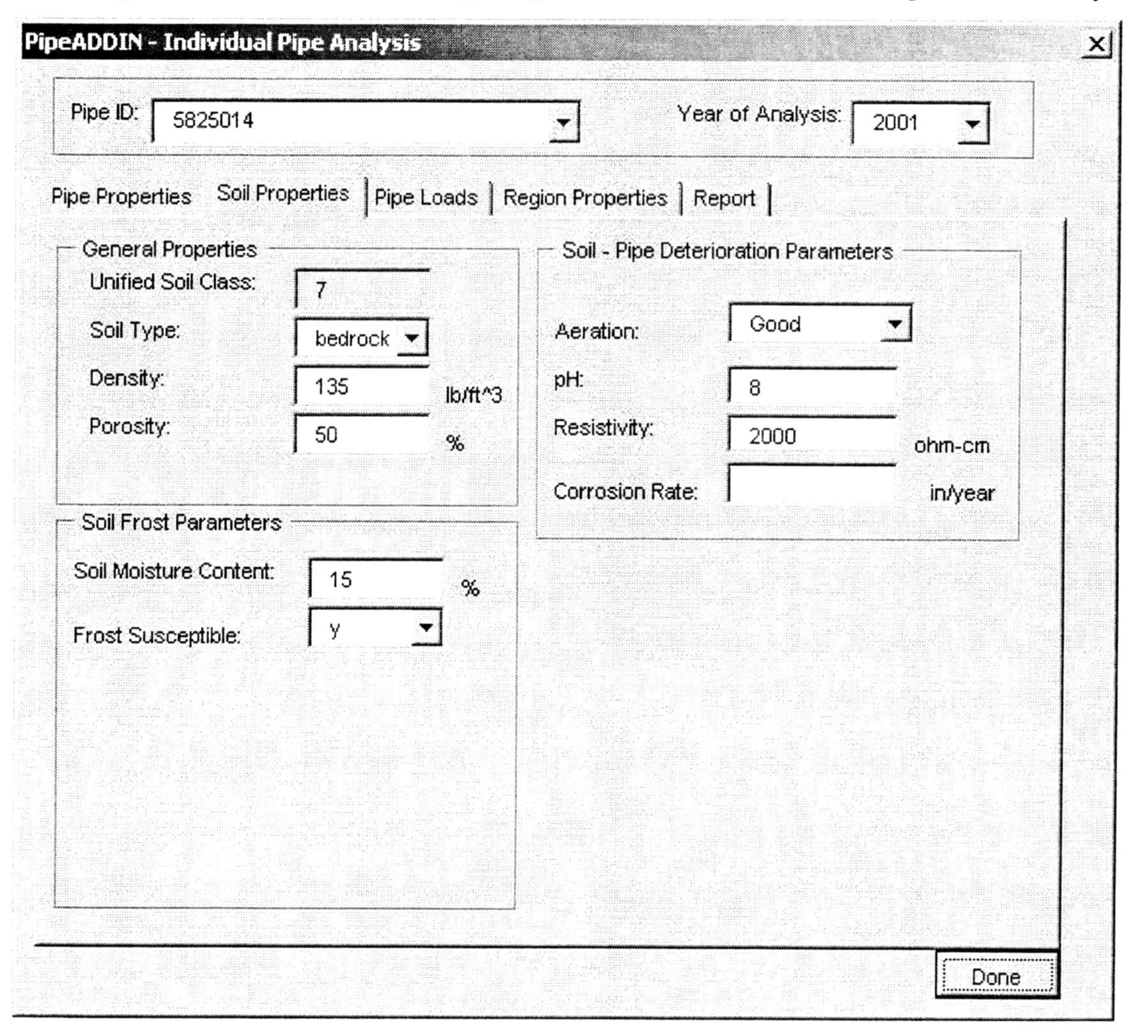

Figure D.14 Soil properties for individual pipe mode

Pipe Loads Form

The default values for each pipe on this form are loaded when the pipe ID changes. The Earth Load, Traffic Load, and Frost Load are all computed based on the pipe properties contained in the BrkInventory Sheet and the soil properties contained in the soils sheet. The user can then change a specific value before continuing with the analysis.

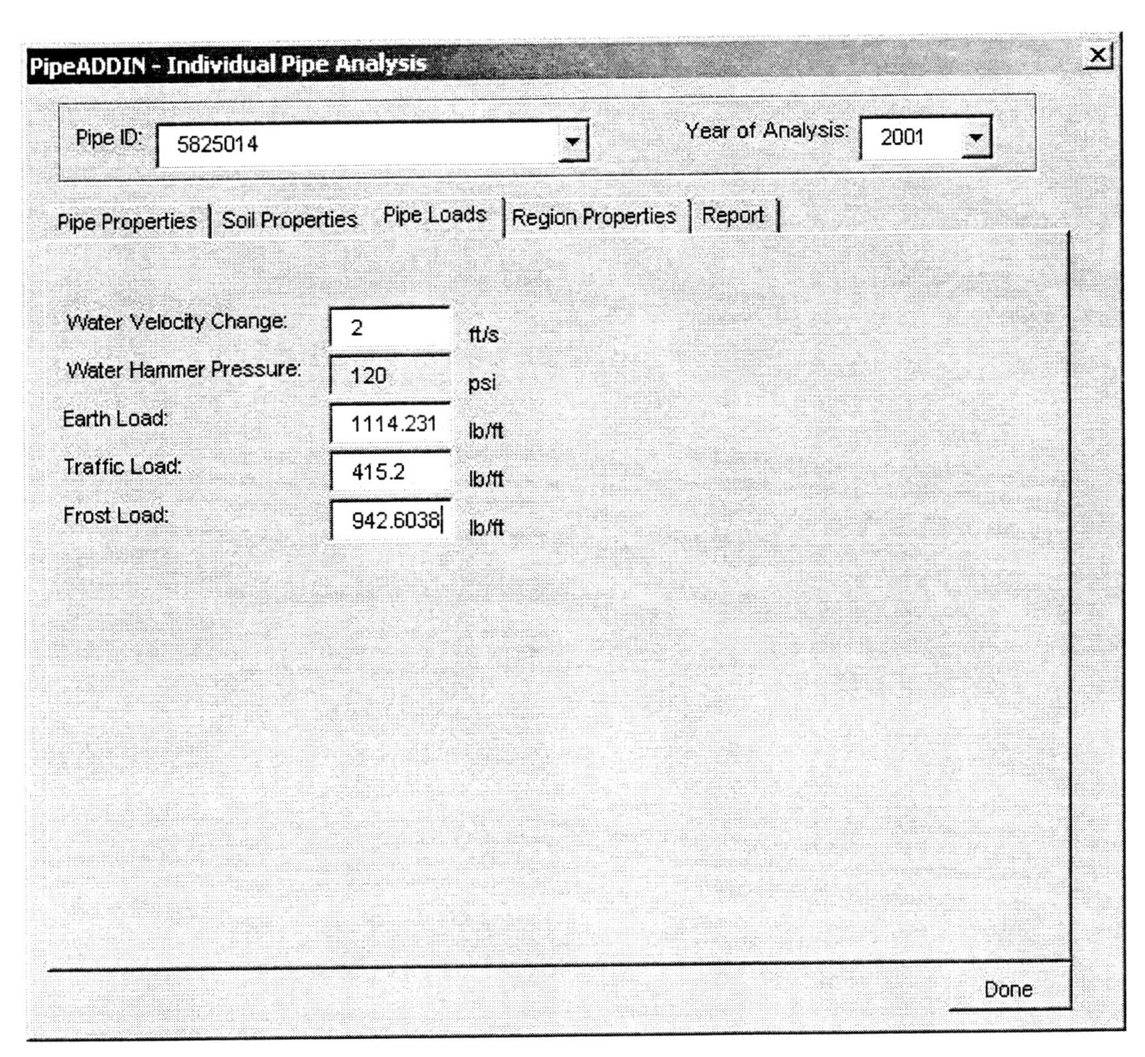

Figure D.15 Pipe loads for individual pipe mode

Region Properties Form

The default values for each pipe on this form are loaded from the 'Regions' Sheet when the pipe ID changes. The values are based on the corresponding values of the reionID for the particular pipe. The user can then change a specific value before continuing with the analysis.

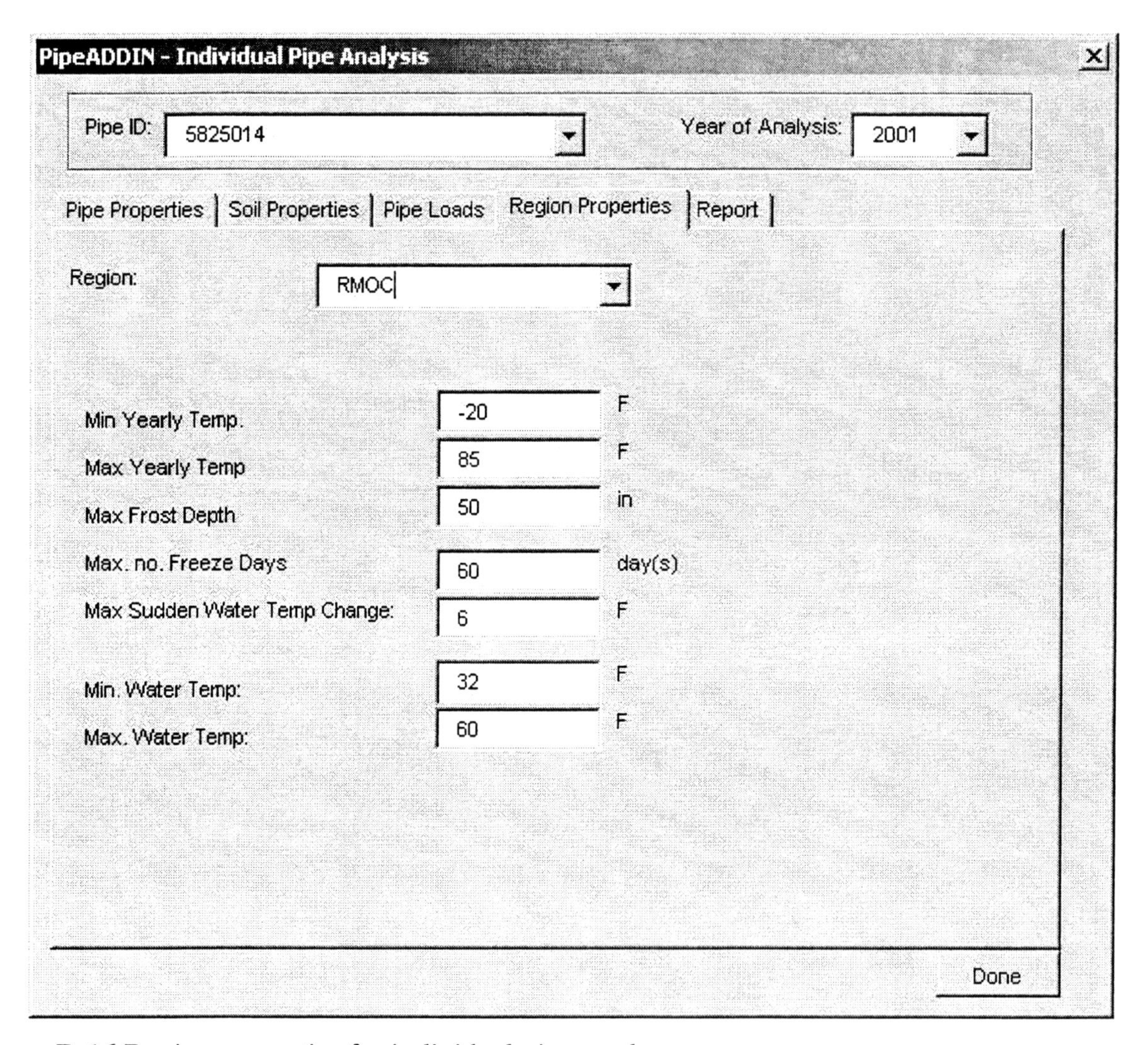

Figure D.16 Region properties for individual pipe mode

Report Form

The result from the individual analysis are displayed in the textbox on this form. Additional results are also displayed on the sheets named 'Summary' and 'ResultsIndividualPipe'.

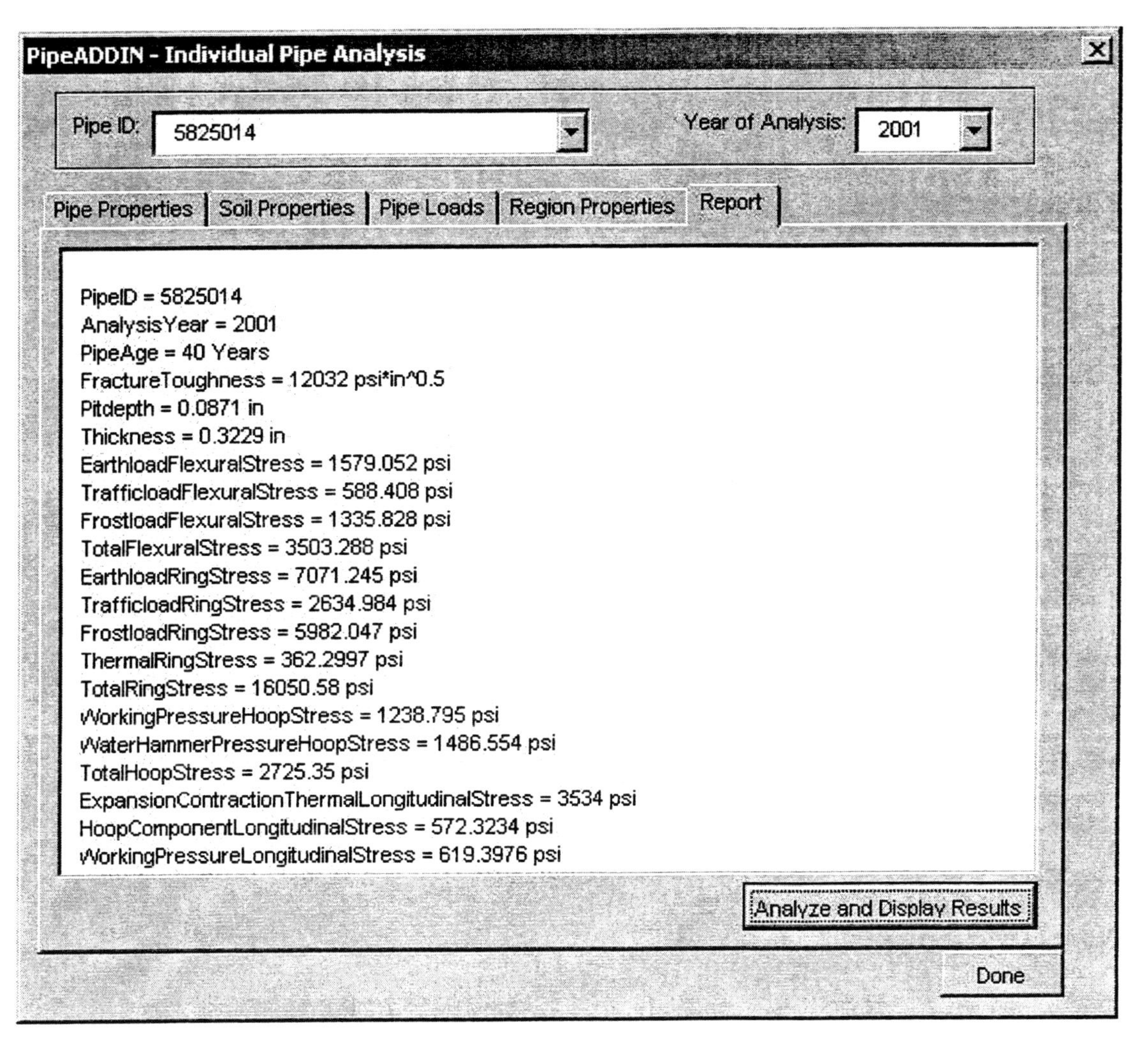

Figure D.17 Output report for individual pipe mode

data.orig.xls

	A	B	C	D	E	F	G	H	I
1									
2	**Conditions at Year 0**								
3		**Pipe ID**	**Year**	**Age, yrs**	**Size, in**	**Length, ft**	**Pitdepth, in**	**Thickness, in**	
4	**Pipe Properti**	5825014	1961	0	8	180.4	0.000	0.410	
5									
6		Load	Pressure	Ring	Hoop	Flexural	Longitudinal	F + L	Ring + Hoop
7				Stress	Stress	Stress	Stress	Stress	Stress
8		lb/L.F.	psi	psi	psi	psi	psi	psi	psi
9	Earth	1114.20		4431.70		1228.55			
10	Traffic	415.20		1651.40		457.80			
11	Frost	942.60		3749.08		1039.32			
12	Internal Pressure		100.00		975.61		487.80		
13	Thermal Load on Pipe		6.00	364.64			341.68		
14	Water Hammer		120.00		1170.73		585.37		
15	Longitudinal Component of Hoop Stress						450.73		
16	Expansion/Contraction Thermal Stress						3534.00		
17	Total Stress	2472.00		10196.83	2146.34	2725.67	5399.59	8125.26	
18	Yield Strength (F)			45000.00	21000.00	27426.00	30000.00	27426	
19	Safety Factor (F)			4.41	9.78	10.06	5.56	3.38	3.41
20	Minimum Safety Factor (F)								3.38

BrkInventory / Soils / Regions / Summary / ResultsIndividualPipe / ResultsAtZeroYea

Figure D.18 Formatted output on Excel sheet for individual pipe mode (Year 0)

data.orig.xls

	A	B	C	D	E	F	G	H	I
23									
24	Conditions at Year 40								
25		Pipe ID	Year	Age, yrs	Size, in	Length, ft	Pitdepth, in	Thickness, in	
26	Pipe Properti	5825014	2001	40	8	180.4	0.087	0.323	
27									
28		Load	Pressure	Ring	Hoop	Flexural	Longitudinal	F + L	Ring + Hoop
29				Stress	Stress	Stress	Stress	Stress	Stress
30		lb/L.F.	psi	psi	psi	psi	psi	psi	psi
31	Earth	1114.20		7071.25		1579.05			
32	Traffic	415.20		2634.98		588.41			
33	Frost	942.60		5982.05		1335.83			
34	Internal Pressure		100.00		1238.80		619.40		
35	Thermal Load on Pipe		6.00	362.30			344.03		
36	Water Hammer		120.00		1486.55		743.28		
37	Longitudinal Component of Hoop Stress						572.32		
38	Expansion/Contraction Thermal Stress						3534.00		
39	Total Stress	2472.00		16050.58	2725.35	3503.29	5813.03	9316.32	
40	Yield Strength (F)			19100.57	14939.22	16443.90	16945.70	16443.9	
41	Safety Factor (F)			1.19	5.48	4.69	2.92	1.77	0.92
42	Minimum Safety Factor (F)								0.92

BrkInventory / Soils / Regions / **Summary** / ResultsIndividualPipe / ResultsAtZeroYea

Figure D.19 Formatted output on Excel sheet for individual pipe mode (Analysis Year)

data.orig.xls

	A	B	C	D	E	F	G	H
1	PipeID	AnalysisYı	PipeAge	FractureTo	Pitdepth	Thickness	EarthloadF	TrafficloadF
2	5825014	1961	0	12032	0	0.41	1228.554	457.8008
3	5825014	1962	1	12032	0	0.41	1228.554	457.8008
4	5825014	1963	2	12032	0.0127	0.3973	1270.097	473.281
5	5825014	1964	3	12032	0.0208	0.3892	1298.039	483.693
6	5825014	1965	4	12032	0.0269	0.3831	1319.767	491.7899
7	5825014	1966	5	12032	0.0318	0.3782	1337.834	498.522
8	5825014	1967	6	12032	0.036	0.374	1353.451	504.3413
9	5825014	1968	7	12032	0.0396	0.3704	1367.298	509.5013
10	5825014	1969	8	12032	0.0427	0.3673	1379.799	514.1597

ResultsIndividualPipe / ResultsAtZeroYear / ResultsAtAnalysisYear / Tem

Figure D.20 Year by year output for individual pipe

Multiple Pipe Mode

In the multiple pipe mode, several pipes can be analyzed together. This is useful when making generalizations about pipes in the system. In order to analyze pipes in multiple mode, all the pipes and their properties must be entered in the 'BrkInventory' sheet. To use to multiple mode, select 'Multiple Analysis' from the 'PipeADDIN' Menu (see Figure D.21). The results are printed on two sheets 'ResultsAtZeroYear' and 'ResultsAtAnalysisYear'. The sheet 'ResultsAtZeroYear' contains the conditions of all the pipes at the time they were installed. The sheet 'ResultsAtAnalysisYear' contains the conditions of all the pipes at the analysis year.

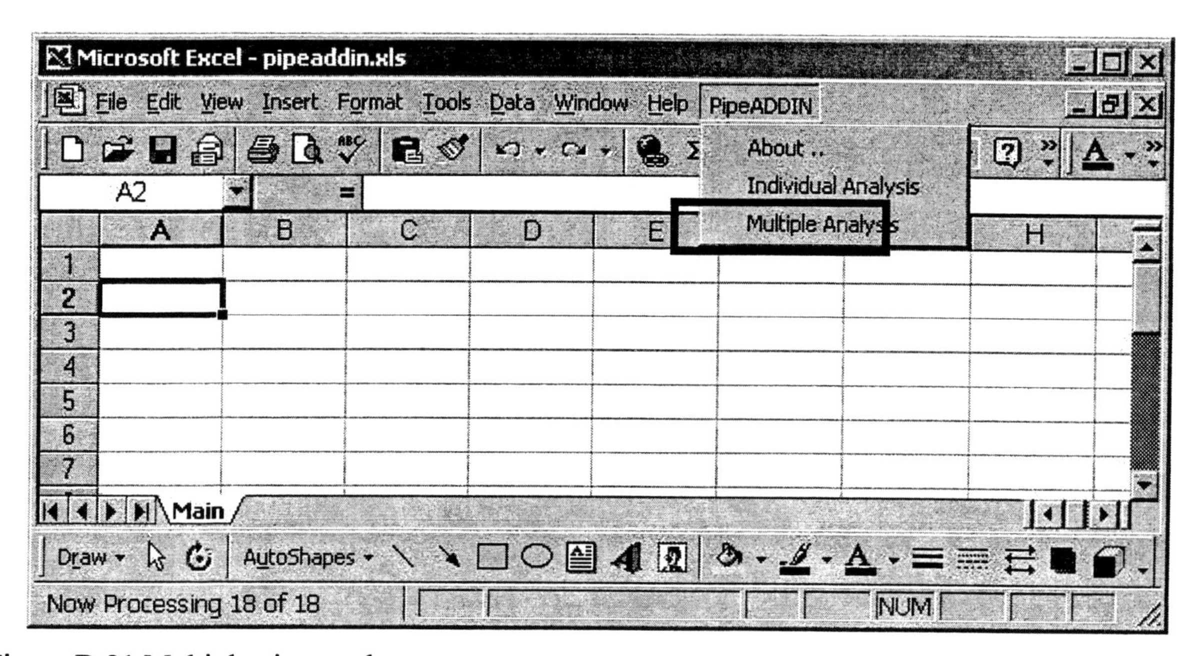

Figure D.21 Multiple pipe mode menu

Figure D.22 shows a pipe by pipe output at time of installation (Year 0) resulting from running the software in multiple pipe mode.

data.orig.xls

	A	B	C	D	E	F	G	H
1	PipeID	Earthload	Trafficload	Frostload	TotalVertic	FractureTo	Pitdepth	Thickness
2	5825014	1114.2	415.2	942.6	2472	12032	0	0.41
3	6021130	1034.6	80.2	374.7	1489.5	12032	0	0.38
4	6224067	1034.6	0	374.7	1409.3	12032	0	0.38
5	6226013	878.9	168.1	703.8	1750.9	12032	0	0.38
6	6226171	1072.4	182.8	485.1	1740.3	12032	0	0.38
7	6426099	927.3	158.6	635.7	1721.6	12032	0	0.38
8	6428276	588.7	438.2	1133	2159.9	12032	0	0.28
9	6624027	766.7	119	654.4	1540	12032	0	0.28
10	6625126	766.7	0	654.4	1421	12032	0	0.28

Soils / Regions / Summary / ResultsIndividualPipe / ResultsAtZeroYear

Figure D.22 Pipe by pipe output for multiple pipe analysis (Year 0)

Figure D.23 shows a pipe by pipe output at year of analysis resulting from running the software in multiple pipe mode.

data.orig.xls

	A	B	C	D	E	F	G	H
1	PipeID	Earthload	Trafficload	Frostload	TotalVertic	FractureTo	Pitdepth	Thicknes
2	6624027	766.7	119	654.4	1540	12032	0.1739	0.106
3	6625126	766.7	0	654.4	1421	12032	0.1709	0.109
4	6428276	588.7	438.2	1133	2159.9	12032	0.1051	0.174
5	6826036	588.1	390.7	930.9	1909.7	12032	0.1541	0.225
6	7031059	733.8	309.1	966.8	2009.7	12032	0.0767	0.203
7	7033002	975.7	99.3	577.5	1652.5	12032	0.0817	0.198
8	8037127	787.7	139.4	352.9	1280.1	12032	0.1757	0.204
9	6825129	1302.5	84.6	255	1642.1	12032	0.1427	0.237
10	5825014	1114.2	415.2	942.6	2472	12032	0.0838	0.326

ResultsIndividualPipe / ResultsAtZeroYear / ResultsAtAnalysisYear / Tem

Figure D.23 Pipe by pipe output for multiple pipe analysis (Analysis Year)

REFERENCES

Ahammed, M. and R.E. Melchers. 1994. Reliability of Underground Pipelines Subject to Corrosion. *Jour. Transportation Engineering*. 120(6):989-1002.

Andreou, S.A., D.H. Marks, and R.M. Clark. 1987. A New Methodology for Modelling Break Failure Patterns in Deteriorating Water Distribution Systems. *Advances in Water Resources*. 10:11-20.

AWWA. 1999. *Mainstream*. 43(2):3.

AWWA. 1998. *WATER:\STATS, The Water Utility Data Base*. Denver, Colo.:AWWA.

Basalo, C. 1992. *Water and Gas Mains Corrosion, Degradation and Protection*. Chichester, England: Ellis Horwood.

Brady, N.C. 1990. *The Nature and Properties of Soils*. 10th ed. New York: Macmillan Publishing Company.

CIPRA (Cast Iron Pipe Research Association). 1978. *Handbook: Ductile Iron Pipe and Cast Iron Pipe*. Fifth Edition.

Clark, R.M., C.L. Stafford, and J.A. Goodrich. 1982. Water Distribution Systems: A Spatial and Cost Evaluation. ASCE *Jour. Water Resources Planning and Management*. 108(3):243-256.

Cohen, A., and M.B. Fielding. 1979. Prediction of Frost Depth: Protecting Underground Pipes. *Jour. AWWA*. 71(2):113-116.

Computer Technology Institute (CTI). 1997. Reliability-Based Decision Support System for the Maintenance Management of the Underground Networks of Utilities. Publishable Synthesis Report BE 7120. Patras, Greece.

Deb, A.K., K.A. Momberger, Y.J. Hasit, and F.M. Grablutz. 2000. *Guidance for Management of Distribution System Operation and Maintenance*. Denver, Colo.: AwwaRF and AWWA, forthcoming.

Deb, A.K., Y.J. Hasit, F.M. Grablutz, and R.K. Herz. 1998. *Quantifying Future Rehabilitation and Replacement Needs of Water Mains*. Denver, Colo.: AwwaRF and AWWA.

Deb, A.K., Y.J. Hasit, and F. Grablutz. 1995. *Distribution System Performance Evaluation*. Denver, Colo.: AwwaRF and AWWA.

DIPRA (Ductile Iron Pipe Research Association). 1984.Handbook of Ductile Iron Pipe and Cast Iron Pipe. 6th Edition. Ductile Iron Pipe Research Association. Birmingham, Alabama.

Duan, N.L., L.W. Mays, and K.E. Lansey. 1990. Optimal Reliability based Design of Pumping and Distribution Systems. ASCE *Jour. Hydraulic Engineering*. 116(2):249-268.

Fielding, M.B., and A. Cohen. 1988. Prediction of Pipeline Frost Load. *Jour. AWWA*. 80(11):62-69.

Flinn, R.A., and P.K. Trojan. 1990. *Engineering Materials and Their Applications*. Boston, Mass.: Houghton Mifflin Co.

Goulter, I.C., and A. Kazemi. 1988. Analysis of Water Distribution Pipe Failure Types in Winnipeg, Canada. *Jour. Transportation Engineering*. 15:.91-97.

Goulter, I.C., and A. Kazemi. 1989. Spatial and temporal groupings of water main breakage in Winnipeg. *Canadian Jour. Civil Engineering*. 115(2):95-111.

Grablutz, F. and S. Hanneken, 2000. Economic Modeling for Prioritizing Pipe Replacement Programs. Presented at the AWWA Infrastructure Conference and Exhibition. Baltimore, MD, 14 March 2000.

Issa, K. 1997. Influence of Moisture on Soil Swelling. *Soil Mechanics and Foundation Engineering*. 34(5):139-144.

Habibian, A. 1994. Effect of Temperature Changes on Water Main Breaks. ASCE *Jour. Transportation Engineering*. 120(2):312-321.

Kane, M. 1996. Water mains rehabilitation. 210-217.

Karaa, F.A., D.H. Marks, and R.M. Clark. 1987. Budgeting of Water Distribution Improvement Projects. ASCE *Jour. Water Resources Planning and Management*. 113(3):378-391.

Kiefner, J.F. and W.G. Morris. 1997. A Risk Management Tool for a Natural Gas Pipeline Operator. Presented at The Pipeline Risk Management and Reliability Conference. Houston, TX. November 1997.

Kleiner, Y., B.J. Adams, and J. S. Rogers. 1998a. Long-Term Planning Methodology for Water Distribution System Rehabilitation. *Water Resour. Res*. 34: 2039-2051.

Kleiner, Y., B. J. Adams, and J. S. Rogers. 1998b. Selecting and Scheduling Rehabilitation Alternatives. *Water Resour. Res*. 34:2053-2061.

Kumar, A., M. Bergerhouse, and M. Blyth. 1987. Implementation of a Pipe Corrosion Management System. Paper No. 312. *Corrosion '87*. San Francisco, California.

Kumar, A., W. Riggs, and M. Blyth. 1986. Demonstration of the Pipe Corrosion Management System, Corrosion Mitigation and Management System. *Technical Report M-86/08*. US Army Corps of Engineers, Construction Engineering Research Laboratory.

Kumar, A., E. Meronyk, and E. Segan. 1984. Development of Concepts for Corrosion Assessment and Evaluation of Underground Pipelines. *Technical Report M-337*. US Army Corps of Engineers, Construction Engineering Research Laboratory.

Lambe, T.W., and R.V. Whitman. 1969. *Soil Mechanics.* New York: John Wiley.

Lane, P.H. and N.L. Buehring. 1978. Establishing Priorities for Replacement of Distribution Facilities. *Jour. AWWA.* 70:7:355-357.

Lansey, K. E., and L.W. Mays. 1989. Optimization Model Design of Water Distribution System Design. ASCE *Journal of Hydraulic Engineering.*115(10):1401-1418.

Loganathan, G.V., H.D. Sherali, and M.P. Shah. 1990. A Two-Phase Network Design Heuristic for the Minimum Cost Water Distribution Systems Under Reliability Constraints. *Engineering Optimization.* 15:311-336.

Loganathan, G.V., J.J. Greene, and T. Ahn. 1995. A Design Heuristic for Globally Minimum Cost Water Distribution Systems. ASCE *Jour. Water Resources Planning and Management.* 121(2):182-192.

Male, J.W., T.M. Walski, and A.H. Slutsky. 1990. Analyzing Water Main Replacement Policies. ASCE *Jour. Water Resources Planning and Management.* 116(3):362-374.

Mavin, K. 1996. Predicting the Failure Performance of Individual Water Mains. Research Report No. 114. Urban Water Research Association of Australia. ISBN 1876088176.

Mays, L.W. Ed. 1989. *Reliability Analysis of Water Distribution Systems.* New York, N.Y.: ASCE.

Millette, L. and D.S. Mavinic. 1988. The Effect of pH on the Internal Corrosion Rate of Residential Cast-Iron and Copper Water Distribution Pipes. *Canadian Journal of Civil Engineering.* 15, 79-90.

Monie, W.D. and C.M. Clark. 1974. Loads On Underground Pipe Due to Frost Penetration. *Jour. AWWA.* 66:6:353-358.

O'Day, K.D., R. Weiss, S. Chiavari, and D. Blair. 1986. *Water Main Evaluation for Rehabilitation /Replacement*. Denver, Colo.: AwwaRF and AWWA.

Park, H., and J.C. Liebman. 1993. Redundancy-Constrained Minimum-Cost Design of Water Distribution Nets. ASCE *Jour. Water Resources Planning and Management*. 119(1):83-98.

Philadelphia Water Department (PWD). 1985a. *Water Main Structural Condition Assessment Model*. Philadelphia, Pa.

Philadelphia Water Department (PWD). 1985b. *Water Main Pipe Sample Physical Testing Program*. Philadelphia, Pa.

PPK Consultants Pty Ltd. 1993. *Identification of Critical Water Supply Assets*. Research Report No. 57. Urban Water Research Association of Australia.

Rajani, B., and C. Zahn. 1996. On the estimation of frost load. *Canadian Geotechnical Jour.* 33(4):629-641.

Rajani, B., J. Makar, S. McDonald, C. Zhan, S. Kuraoka, C.K. Jen, and M. Viens. 2000. *Investigation of Grey Cast Iron Water Mains to Develop a Methodology for Estimating Service Life*. Denver, Colo.: AwwaRF and AWWA.

Romanoff, M. 1957. Underground Corrosion. *National Bureau of Standards Circular 579*. US Government Printing Office.

Rossum, J.R. 1969. Predicting of Pitting Rates in Ferrous Metals from Soil Parameters. *Jour. AWWA*. 61:6, pp.305-310.

Roy F. Weston, Inc. and TerraStat Consulting Group. 1996. Pipe Evaluation System (PIPES): Deterioration Model Statistical Analysis. Unpublished Project Report.

Shamir, U., and C.D.D. Howard. 1979. An Analytic Approach to Scheduling Pipe Replacement. *Jour. AWWA.*

Sherali, H.D., R. Totlani, and G.V. Loganathan. 1998. Enhanced Lower Bounds for the Global Optimization of Water Distribution Networks. *Water Resources Research.* 34(7):1831-1841.

Smith, R. A. 1977. "A Simplified Method of Predicting the Rates of Growth of Cracks Initiated at Notches". Fracture Mechanics in Engineering Practice . P. Stanley, Ed. Applied Science, London pp.173-182

Spangler, M.G., and R.L. Handy. 1982. *Soil Engineering.* 4th ed. New York: Harper and Row Publishers.

Stacha, J.H. 1978. Criteria for Pipeline Replacement. *Jour. AWWA.* 70(5):256-258.

Su, Y.C., L.W. Mays, and K.E. Lansey. 1987. Reliability Based Optimization Model for Water Distribution Systems. ASCE *Jour. of Hydraulic Engineering.* 114(12):589-596.

Sullivan, J.P., Jr. 1982. Maintaining Aging Systems-Boston's Approach. *Jour. AWWA.*

Timoshenko, S. and J. N. Goodier. 1987. Theory of Elasticity. McGraw-Hill, New York New York,

USEPA (U.S. Environmental Protection Agency). 1997. *Drinking Water Infrastructure Needs Survey. First Report to Congress.* EPA-812-R-97-001. Washington, DC. USEPA.

U.S. Pipe and Foundry Co. 1962. An Attempt at Explaining Why Pipe Failures Increase under Certain Winter Conditions. Memorandum of Record. March 19, 1962.

Wagner, J.M., U. Shamir, and D.H. Marks. 1988. Water Distribution Reliability. ASCE *Jour. Water Resources Planning and Management.* 114(3):276-294.

Walski, T.M. 1984. *Analysis of Water Distribution Systems.* New York, NY. Van Nostrand Reinhold Co.

Walski, T.M. 1987. *Replacement Rules for Water Mains. Jour. AWWA.* 79(11):33-37.

Walski, T.M. and A. Pellicia. 1982. Economic Analysis of Water Main Breaks. *Jour. AWWA.* 74(3):40-147.

Wedge, R .F. 1990. Evaluating Thermal Pressure Changes in Liquid Packed Piping. In *Pipeline Design and Installation.* Edited by K.K. Kienow. New York, NY. ASCE.

ABBREVIATIONS

A	a constant in the corrosion pit equation (= 0.22 for cast iron pipe)
a	velocity of pressure wave
a	measure of crack length
a	distance from center to inner wall
A	area of pipe exposed to the soil
ASTM	American Society for Testing and Materials
AWWA	American Water Works Association
AwwaRF	Awwa Research Foundation
α	linear expansive coefficient of the pipe material
b	distance from center to outer wall
B_d	width of trench at top of pipe
BS&J	Buck, Seifort & Jost, Inc.
C	repair cost of a break
C	surface load factor for unpaved of flexible pavement
C_d	calculation coefficient
CI	cement lined cast iron
CIPRA	Cast Iron Pipe Research Association
COC	chain-of-custody
CSI	corrosion status index
CTI	Computer Technology Institute
d	nominal inside pipe diameter
D	outside diameter of the pipe
DBMS	database management system
DPA	deterioration point assignment
DSS	decision support system
ΔT	temperature difference

E	modulus of elasticity
ES	expert system
°F	degrees Fahrenheit
F	replacement cost per unit length of pipe
F	Impact factor
Fd_{max}	maximum frost depth supplied by the user
Fd_c	frost depth computed from the Rajani and Zahn equation
ft	foot, feet
ft^2	feet square
FWWD	Ft. Worth Water Department
g	acceleration of gravity
GIS	geographic information system
GPS	global positioning system
GRI	Gas Research Institute
γ	unit weight of soils overlying pipe
H	height of fill above the top of pipe
in.	inch
k	fluid bulk modulus
K	surface load factor for rigid pavement
K_a	a constant in the corrosion pit equation (= 1.40 for cast iron pipe)
K_n	a constant in the corrosion equation describing soil type
K_r	ratio of active horizontal pressure to the vertical pressure
K_0	fracture toughness of the material
K_{IC}	fracture toughness

L	length of pipe
L	length of unsupported span
LADWP	Los Angeles Department of Water and Power
LNEC	Laboratorio Nacional de Engenharia Civil (Civil Engineering National Laboratory)
LWC	Louisville Water Company
MDPE	medium density polyethylene
mgd	million gallons per day
mi	mile
mld	million litres per day
MTL	Materials Testing Laboratory
μ'	coefficient of sliding friction between fill materials and sides of trench
n	a constant in the corrosion equation describing soil type
NBS	National Bureau of Standards
NRC	National Research Council
ν	Poisson's ratio
O&M	operation and maintenance
ω	soil resistivity
p	pit depth
p	internal pressure
P	surge pressure
P	wheel load
P	internal pressure at bursting for the pipe
P	water pressure
P	bursting pressure
P_{sw}	swelling pressure
PAC	Project Advisory Committee
PBM	Pipe Break Model

PDM	Pipe Deterioration Model
PEM	Pipe Evaluation Model
PIPES	Pipe Evaluation System
PLM	Pipe Load Model
psi	pounds per square inch
psf	pounds per square foot
PSWC	Philadelphia Suburban Water Company
PWD	Philadelphia Water Department
r	distance of any point from the center
r_n	radius of the root of the notch
r_n	depth of the corrosion pit
R	discount rate
R	reduction factor, accounting for the part of the pipe directly below the wheels
RMOC	Regional Municipality of Ottawa-Carleton
RMT	risk management tool
SCC	stress corrosion cracking
SCM	Statistical Correlation Model
SLCWC	St. Louis County Water Company
SF_f	safety factor due to flexural stresses
$SF_{flexlong}$	safety factor due to flexural plus longitudinal stresses
SF_h	safety factor due to hoop stresses
SF_l	safety factor due to longitudinal stresses
SF_θ	safety factor due to ring stresses
σ	nominal stress at fracture
σ_f	sum of flexural stresses
$\sigma_{flexlong}$	sum of flexural and longitudinal stresses
σ_h	sum of hoop stresses
σ_l	sum of longitudinal stresses
$\sigma_{l,T}$	longitudinal stress due to thermal contraction

σ_{nom}	smallest stress level to cause complete failure
σ_{res}	residual strength for each of strength category
$\sigma_{res(bts)}$	residual bursting tensile strength
$\sigma_{res(mr)}$	residual modulus of rupture
$\sigma_{res(rm)}$	residual ring modulus of rupture
$\sigma_{res(ts)}$	residual tensile strength
σ_0	plain fatigue strength of the material
σ_θ	sum of ring stresses
σ_v	longitudinal component of hoop stress
t	net thickness
t	pipe wall thickness
T	time of exposure
UCI	unlined cast iron
U.K.	United Kingdom
U.S.	United States
USEPA	United States Environmental Protection Agency
UWNJ	United Water New Jersey
VA Tech	Virginia Polytechnic Institute and State University
V	maximum velocity change
w	external load
W	crushing load
W_e	earth load
W_{fadj}	the adjusted frost load, lb/ft
W_{frz}	the frost load computed from the Rajani and Zahn equation, lb/ft
W_L	liquid limit
W_0	initial soil water content
W_t	truck superload

WESTON	Roy F. Weston, Inc.
WMS	work management system
WSSC	Washington Suburban Sanitary Commission
y	depth of the notch
Y	correction factor that accounts for the geometry of the flawed component
yr	year